TECHNOLOGY ROADMAP FOR SUSTAINABLE WASTEWATER TREATMENT PLANTS IN A CARBON-CONSTRAINED WORLD

by:
George V. Crawford, P.Eng.
CH2M HILL Canada Limited

2010

WERF

The Water Environment Research Foundation, a not-for-profit organization, funds and manages water quality research for its subscribers through a diverse public-private partnership between municipal utilities, corporations, academia, industry, and the federal government. WERF subscribers include municipal and regional water and wastewater utilities, industrial corporations, environmental engineering firms, and others that share a commitment to cost-effective water quality solutions. WERF is dedicated to advancing science and technology addressing water quality issues as they impact water resources, the atmosphere, the lands, and quality of life.

For more information, contact:
Water Environment Research Foundation
635 Slaters Lane, Suite G-110
Alexandria, VA 22314-1177
Tel: (571) 384-2100
Fax: (703) 299-0742
www.werf.org
werf@werf.org

This report was co-published by the following organization.

IWA Publishing
Alliance House, 12 Caxton Street
London SW1H 0QS, United Kingdom
Tel: +44 (0) 20 7654 5500
Fax: +44 (0) 20 7654 5555
www.iwapublishing.com
publications@iwap.co.uk

IWAP ISBN: 978-1-84339-396-2/1-84339-396-4

CH2M HILL Canada Limited

The research on which this report is based was developed, in part, by the United States Environmental Protection Agency (EPA) through Cooperative Agreement No. CR83155901-2 with the Water Environment Research Foundation (WERF). However, the views expressed in this document are not necessarily those of the EPA and EPA does not endorse any products or commercial services mentioned in this publication. This report is a publication of WERF, not EPA. Funds awarded under the Cooperative Agreement cited above were not used for editorial services, reproduction, printing, or distribution.

ACKNOWLEDGMENTS

WERF thanks the volunteer efforts of the Optimization Challenge Interest Area Team, particularly the Technical Review Committee, for the support they gave to the preparation of this report. This report would not be possible without the generous support and contribution of the expert workgroup. Research for this report was conducted under WERF contract OWSO4R07 Task Order 4.

Research Team

Principal Investigator:
George V. Crawford, P.Eng.
CH2M HILL Canada Limited

Project Team:
Julian Sandino, Ph.D., P.E. *CH2M HILL*
Collin Moore, *CH2M HILL*

Technical Review Committee

David Cooley, *Hampton Roads Sanitation District*
Robert Kelly, *Suez Environnement*
David Tucker, *City of San Jose*
James Wheeler, P.E., *U.S. EPA*
John Willis, P.E., *Brown and Caldwell*

Work Group Participants

Rob Bauer, *Clean Water Services*
Charles Bott, *Hampton Roads Sanitation District*
Kartik Chandran, *Columbia University*
Christine DeBarbadillo, *Black & Veatch*
John Dyson, *Degremont Technologies-Infilco*
Eugenio Geraldo, *American Water*
Mark Laquidara, *AECOM*
Nancy Love, *University of Michigan*
Sudhir Murthy, *District of Columbia Water and Sewer Authority (DCWASA)*
Richard Novak, *Praxair*
Krishna Pagilla, *Illinois Institute of Technology*
Denny Parker, *Brown and Caldwell*
Andrew Randall, *University of Central Florida*
Dave Reardon, *HDR*
David Stensel, *University of Washington*
Jeff Weiss, *Los Angeles County Sanitation District*
Tom Wilson, *Consultant*
Heng Zhang, *Metro Water Reclamation Districts of Greater Chicago*
Joe Zuback, *Global Water Advisors (WERF Board of Directors)*

Stakeholder Representatives:
Amit Pramanik, *WERF Senior Program Director – Wastewater*
Robert K. Bastian, *U.S. EPA*
Matt Ries, *Water Environment Federation*
Cora Uijterlinde, *Stichting Toegepast Onderzoek Waterbeheer (STOWA) and Global Water Research Coalition*
Cao Yeshi, *Technology and Water Quality Office, PUB Singapore and Global Water Research Coalition*
Tan Soon Keat, *School of Civil and Environmental Engineering, Nanyang Technological University*
Chua Lian Heong, *DHI-NTU Water & Environmental Research Centre and Education Hub, Singapore*
Mark van Loosdrecht, *Delft University of Technology*
Joseph Sobanksi, *Metro Water Reclamation Districts of Greater Chicago*
Thomas Kunetz, *Metro Water Reclamation Districts of Greater Chicago*

WERF Optimization Challenge Issue Area Team

John Barber, Ph.D., *Eastman Chemical*
Shahid Chaudhry, *California Energy Commissions*
Steve Constable, *DuPont Engineering Research & Technology*
David Cooley, *Hampton Roads Sanitation District*
Glen Daigger, Ph.D., P.E. *CH2M HILL*
Robert F. Kelly, *Degremont Technologies-Infilco*
Melanie S. Masek, P.E. *DuPont Engineering Research & Technology*
Arthur J. Meyers, Jr., Ph.D., *Retired*
Hsiao-Ting Ong, P.E., *City of San Jose*
Ali Oskouie, Ph.D., *Metro Water Reclamation Districts of Greater Chicago*
Steven M. Rogowski, P.E., *Metro Wastewater Reclamation District of Denver*
Gary Shimp, *Black & Veatch*
Rob Simm, *Stantec Consulting, Inc.*
David Tucker, *City of San Jose*
Dariush Vosooghi, Ph.D., P.E., *City of Los Angeles*
Milind Wable, Ph.D., P.E., *CDM*
James Wheeler,P.E. *U.S. EPA*
John Willis, P.E., *Brown and Caldwell*

Water Environment Research Foundation Staff

Director of Research: Daniel M. Woltering, Ph.D.
Program Director: Lauren Fillmore, M.Sc.

ABSTRACT AND BENEFITS

Wastewater contains reusable water, carbon (energy) and nutrients (nitrogen, phosphorus, and sulfur) that could be recovered or reused. Meanwhile, current treatment objectives are to produce an acceptable quality of water for reuse or discharge at the lowest LCC. Most of the current treatment processes manage carbon and nutrients as wastes to be removed, and do not attempt to capitalize on these resources inherent in wastewater. In the context of sustainability and climate change, the next generation of wastewater treatment processes should focus on resource recovery (water reuse, energy/carbon recovery, and nutrient recovery) as much as they currently do on meeting the treatment objectives. The goal is for treatment of domestic wastewater to have a minimal carbon footprint, and to be 100% self-sustainable with regards to energy, carbon, and nutrients, while achieving a discharge or reuse quality that preserves the quality of the receiving waters.

In May 2009, the Water Environment Research Foundation convened a work group of international experts in the wastewater sector to develop a Wastewater Treatment Technology Roadmap that will identify carbon-conscious routes to sustainable wastewater treatment. The resultant Technology Roadmap, presented in this report, identifies pathways toward sustainable wastewater systems over the next few decades, including various approaches the sector could use over the next 20 to 30 years.

The Technology Roadmap describes the current status of wastewater technologies, projects future treatment quality requirements, identifies research needs, and summarizes ongoing activities to meet the perceived future objectives such as reducing the carbon footprint while achieving lower nutrient levels. Work group participants discussed possible technology concepts that can be reasonably expected to produce actionable results and can be implemented by interested wastewater utilities. The participants considered typical and atypical approaches to optimizing carbon and nutrient management at WWTPs. Typical approaches include the evaluation of process modeling opportunities and constraints, and incremental resource and carbon management optimization techniques.

Atypical approaches will be even more important to the future of wastewater resource reclamation. As an additional outcome, several working group members suggested conceptual and sustainable "plant of the future" treatment systems not constrained by existing infrastructure. The identified concepts can be expected to generate opportunities and research needs related to energy sources within treatment plants, changing wastewater characteristics, decentralized treatment, increased nutrient recovery and management, and total water reuse.

Benefits:

- Anticipates future treatment requirements and research needs
- Summarizes ongoing activities to meet the perceived future objectives such as reducing the carbon footprint while achieving lower nutrient levels.
- Identifies possible technology concepts that can be implemented by interested utilities.

Keywords: Sustainable wastewater treatment, carbon footprint, resource recovery, research, and development.

TABLE OF CONTENTS

LIST OF TABLES

LIST OF FIGURES

LIST OF ACRONYMS

BOD	biochemical oxygen demand
CO_2	carbon dioxide
COD	chemical oxygen demand
CSO	combined sewer overflow
FOG	fats, oils, and grease
GHG	greenhouse gas
I/I	infiltration and inflow
kWh/MG	kilowatt hours per million gallons
LCA	life cycle assessment
LCC	life cycle cost
N	nitrogen
O_2	oxygen
O&M	operations and maintenance
P	phosphorus
R&D	research and development
S	sulfur
SRT	solids retention time
TBL	triple bottom line
TMDL	total maximum daily load
WERF	Water Environment Research Foundation
WWTP	wastewater treatment plant
UASB	upflow anaerobic sludge blanket
UV	ultraviolet
U.S. EPA	U.S. Environmental Protection Agency
VFA	volatile fatty acids

EXECUTIVE SUMMARY

Resources end up in wastewater through inefficient consumption. As a result, wastewater contains potentially reusable water, carbon (energy), and nutrients (nitrogen, phosphorus, and sulfur) that could be recovered. Meanwhile, current wastewater treatment objectives are primarily to produce an acceptable quality of treated water at the lowest life cycle cost (LCC). Most current treatment processes manage carbon and nutrients as wastes to be removed, and do not attempt to capitalize on the resources inherent in wastewater. In the context of sustainability and climate change, the next generation of wastewater treatment processes should focus on resource recovery (water reuse, energy/carbon recovery, and nutrient recovery) as much as they currently do on treatment. The future goal for wastewater treatment of domestic wastewater should be to have a minimal carbon footprint, and to be 100% self-sustainable with regards to energy, carbon, and nutrients, while achieving a discharge or reuse quality that preserves the quality of the receiving waters.

Such a paradigm shift in the wastewater sector will not occur overnight. Significant changes in water sector practices can be achieved if research is planned to remove obstacles and investigate unknowns. This Technology Roadmap Report defines a vision of a plausible future and predicts the changes in practice needed to meet anticipated requirements in the future. The Technology Roadmap identifies research gaps that must be addressed in a timely fashion to sustain positive change towards resource recovery and the conservation or wise use of resources including embedded energy. The Technology Roadmap identifies possible pathways toward achieving sustainable water reclamation systems over the next few decades, including new and re-vitalized old approaches to wastewater treatment. These changes are for utilities to implement over the next 20 to 30 years. Over two days, a work group of wastewater experts developed the Water Environment Research Foundation (WERF) Technology Roadmap; this report summarizes the findings and recommendations.

Future Effluent and Reuse Quality Targets

Several water quality requirements are expected to drive the development of treatment targets and other key plant design and operating constraints in the future. The experts anticipate more water quality-based discharge limits from designated use-based standards, and permits will be driven by both total maximum daily load (TMDL) and public health perspectives. Future effluent requirements will likely include microconstituents limits, targeted pathogens instead of surrogate indicators, reduction of nutrients to very low concentrations, low levels of metals, and standards for nanoparticles.

Wastewater experts predict the establishment of tiered water reuse criteria (potable, agriculture, industrial, algae growth) with the result that all discharges will be ecologically neutral or enhancing to their intended use. They also expect that residuals produced through wastewater treatment will have tiered reuse criteria for either agriculture or industrial products. Microconstituents will be controlled in residuals for reuse, and residual energy recovery will become common practice. Zero residuals will be the objective for disposal – all should result in marketable products.

Greenhouse gas (GHG) emissions from the domestic wastewater reclamation sector will be limited, with only biogenic carbon leaving the site. Future regulations may limit the mass of GHG released on a per capita basis, and energy consumption metrics, such as kilowatt-hour per million gallons (kWh/MG) or per capita, will become part of the water sector's operational metrics.

Most significantly, the embedded energy in domestic wastewater will be recognized and recovered. The potential energy in wastewater, including from residuals, liquid stream constituents, and as recoverable heat, exceeds the energy to treat it 10-fold; yet currently this renewable potential resource largely goes to waste. As future regulations regarding discharge quality become increasingly stringent, current practices would have facilities using more advanced treatment resulting in increased energy consumption. Without a concerted effort to help wastewater treatment facilities become net energy producers, they will continue with current practices, and will demand substantial energy to operate.

Actionable Technology

Brainstorming sessions and ranking exercises by the expert participants resulted in predictions about processes and technologies that will be needed in the future. The experts believe that communications and public education will play a large role in the implementation of future plant concepts and are needed by the water reclamation sector to promote understanding of our objectives and engage public support. The experts felt that change towards sustainable water reclamation facilities can only be achieved when advances in technology are supported by changes in public's view and the sector's practices. This means that public outreach on the value of reused water and recovered products is critical for the water reclamation sector. The use of comprehensive decision support tools to balance the trade-offs between carbon footprint reduction, energy recovery, nutrient recovery, discharge water quality, and cost in wastewater systems will be commonplace.

Research Gaps

The most interesting and thought-provoking outcome of the Technology Roadmap is the list of research gaps and opportunities to transform the wastewater sector into a resource-neutral and energy self-sufficient industry. The experts identified gaps and opportunities, grouped them into categories, and then ranked them in terms of priority of their expectation to achieve the sustainability goals. Ensuing discussions then consolidated the experts' thoughts, and some highest priority overall themes were identified. The prioritized themes, grouped into categories, are as follows:

Management Category

- Improved metrics for sustainability
- Improved decision support systems and methodologies
- Public education – rebranding from wastewater treatment to reclamation and recovery
- Recognize the research and development (R&D) funding paradigm for new technology development in a public service sector
- Change the sector mind-set toward risk management, rather than risk aversion
- Improved technology transfer – enable research to cross over into applied knowledge
- Augment the feedstock to the treatment plant (additional carbon sources)
- Revised regulatory approach – incentive-based rather than punitive

Treatment Processes Category

- Further reliance on biotechnology
- Optimization of carbon to energy within liquid and solids treatment processes
- Fixed film processes
- Nutrient recovery processes for nitrogen (N), phosphorus (P), and sulfur (S)
- Enhanced primary treatment for increased carbon diversion to solids processing
- Membrane equipment and process improvements to allow broader application
- Microconstituents / micropollutants – fate and effects research
- Develop and use new thermal sludge processes (pyrolysis and gasification)
- Control or capture air and GHG emissions

Equipment Category

- Heat recovery from sidestreams and plant effluent
- Better aeration systems

"Other" Category

- Better disinfection processes and equipment

Technology Pathway

Wastewater treatment operations are essential public services to protect health and the environment. Engineers and operators can be the agents of innovation and help achieve the goals of energy self-sufficiency, nutrient recovery, and GHG reduction while achieving water quality objectives. However, substantial investment in research is necessary. To capture and use resources from wastewater, investments must be made in research over the next 10 years that will:

- Define whole plant treatment concepts that are energy neutral, including multi-sludge systems and sidestream treatment processes;
- Extract nutrients in forms that are amenable for agricultural reuse with minimal perception of risk to public health;
- Improve the use of biogas to produce energy;
- Identify and develop those emerging biosolids-to-energy technologies that show the greatest promise so that utilities can cost-effectively use them; and
- Explore new concepts that extract energy from wastewater, such as microbial fuel cells, or that produce or use other energy byproducts from wastewater such as hydrogen or nitrous oxide.

Potential resources in the form of domestic wastewater are available in every community across America. Investment in research can make renewable energy available, recover nutrients, and provide sustainable, carbon-neutral water reclamation services.

CHAPTER 1.0

TECHNOLOGY ROADMAP OVERVIEW

1.1 Optimization of Wastewater and Solids Operations

In 2005, the Water Environment Research Foundation (WERF) initiated program-directed research, a research approach that attempts to deliver greater value to subscribers by making the results of important research programs more actionable. As a result, one of these programs, the optimization of wastewater and solids operations, commenced in response to subscriber needs. The primary goal is to develop methods and technologies that will allow the wastewater sector to achieve treatment goals while reducing the resources expended by 20% or more. To accomplish these goals, research must be planned and conducted to satisfy not only immediate technology and research needs but also the needs of the plants of the future. Those future plants will be challenged to meet more stringent treatment requirements and higher water reuse demands while recovering energy and resources.

1.2 Background

Resources end up in wastewater through inefficient consumption (including use of garbage disposal, and industrial wastewater discharges). As a result, wastewater contains reusable water, reduced carbon (energy), and nutrients (nitrogen, phosphorus, and sulfur) that could be recovered and/or reused. Most of the current treatment processes do not sufficiently capitalize on the recovery of these resources that are inherent in wastewater. These materials have instead been perceived as a waste to be removed. In the context of sustainability and climate change, the next generation of wastewater treatment processes should focus on resource recovery (water reuse, energy/carbon recovery, and nutrient recovery) as much as they currently do on treatment and cost reduction. The future goal for the treatment of domestic wastewater should be to have a minimal carbon footprint, and to be 100% self-sustainable with regards to energy, carbon, and nutrients, while achieving a discharge or reuse quality that preserves the quality of the receiving waters.

Modern wastewater treatment evolved to protect the public from waterborne disease over a century ago. With the passage of the *Clean Water Act* in 1972, wastewater systems in the United States had to meet required standards, such as providing secondary treatment and meeting water quality standards. By 2004, according to the U.S. Environmental Protection Agency (U.S. EPA) Clean Watershed Needs Survey, there were over 16,500 publicly owned wastewater treatment facilities serving approximately 230 million people. Before the passage of the 1972 *Clean Water Act*, only half of publicly owned treatment facilities treated wastewater to standards at or above secondary treatment [30 mg/l biochemical oxygen demand (BOD) and 30 mg/l total suspended solids]. By 2004, less than 0.2% of these plants serving approximately 1% of the sewered population did not meet or exceed these discharge standards. Approximately 30% of the facilities now provide some level of advanced treatment. With all the success experienced nationwide by the wastewater sector in providing better treatment over the past 40 years, energy efficiency, or energy and resource recovery have not been objectives of wastewater service providers. It is now widely recognized that wastewater treatment has a carbon footprint and that wastewater contains embedded heat and energy that can be recovered or better used to improve treatment process efficiency.

1.3 Vision of Sustainable Recovery from Wastewater

Although energy self-sufficiency, carbon footprint, and resource recovery through wastewater treatment were widely recognized by the experts as being important, there was not an immediate consensus regarding how these elements can be combined into a workable definition of a sustainable plant. Optimizing energy recovery alone, perhaps at the expense of treated water quality, GHG emissions, and/or nutrient recovery, does not make a plant sustainable. More broadly stated the optimum solution for any of the above components will most likely be a sub-optimal solution for the collective component objectives.

It is not possible to transform the above components of sustainability (energy, water quality, GHG, nutrient recovery, etc.) into purely economic or financial analysis; the value of all components including the quality of water produced must be considered, as well as the value of aesthetic features including noise, odor and visual distraction.

A consensus eventually emerged regarding the experts vision of a sustainable plant. First, the plant should be self-sufficient for energy, ideally producing and recovering as much energy from the wastewater and residuals as is required to operate the treatment plant. The plant should minimize its carbon footprint and GHG emissions; the experts recognized the biogenic carbon emission level as a target, however most agreed that a measurable metric was needed, perhaps in terms of a mass of GHG per capita. The water quality must meet or enhance the desired use of the water produced, whether for irrigation, reuse or watershed restoration. Nutrients, particularly nitrogen and phosphorus, must be recovered in marketable form. And the plant facilities and operation must fit in with the surrounding area.

The need for new and improved decision support tools and methodologies was recognized by the experts, to ensure that all the various components are included and evaluated.

1.4 Report Structure

This Technology Roadmap identifies pathways toward sustainable wastewater systems over the next few decades. Interaction, discussion, and participation with foremost experts in wastewater treatment processes arena resulted in the pathways and targets that form the basis of this Technology Roadmap. The Technology Roadmap includes various approaches toward sustainable wastewater treatment that utilities could use with achievable milestones over the 20-30 years. Part of the Technology Roadmap process is to identify technology gaps and non-technology needs that should be addressed systematically. One ingredient of the Technology Roadmap is the identification of appropriate metrics to allow measurement of progress toward selected objectives, whether they are to optimize energy production and use, carbon footprint, resource consumption and recovery, and/or other factors associated with sustainable plant design and operation.

This report will follow the progress of the workshop, reporting the findings as they were developed by the workshop participants. The report will include:

1. Description of the current status of wastewater treatment objectives and technologies,
2. Ongoing activities, and future needs, to meet impending objectives from reducing the carbon footprint to achieving lower nutrient levels, and
3. Conceptual optimized treatment systems not constrained by existing infrastructure.

CHAPTER 2.0

EXPERT WORK GROUP

2.1 Work Group Approach

A working group of experts developed this Wastewater Treatment Technology Roadmap to describe possible routes to sustainable wastewater treatment in the new, carbon-constrained world. A workshop approach was adopted by WERF to bring together wastewater sector leaders in a forum that facilitates creative thinking about future technologies and treatment objectives. The workshop, held in Chicago on May 20 and 21, 2009, was attended by a variety of recognized water industry leaders, representing academia, utilities, equipment manufacturers, regulators, international associations, and consulting firms.

The workshop was structured to focus primarily on four fundamental questions:

1. What effluent or reuse quality will be needed in 2040?
2. What will be the common technologies in 2040 to achieve that quality?
3. Can those technologies be combined to achieve carbon and energy-neutral treatment?
4. What research and breakthroughs are needed to create new technologies or to improve current technologies to make wastewater treatment plants (WWTPs) sustainable and energy self sufficient in the future?

While addressing the above four central questions, the working group discussed key supporting topics, including the definition of a "roadmap," and the boundary conditions for the brainstorming and roadmapping activities.

Some parts of the workshop were conducted using breakout groups to allow greater discussion and a higher level of detail to be generated on specific topics. This approach had several advantages, such as creating an initial ranking of research needs based on how many breakout groups identified that particular need during their breakout session.

2.2 Purpose of the Workshop

The purpose of the workshop was to brainstorm all possible technology concepts provided they can be reasonably expected to produce actionable results that can be implemented by interested wastewater utilities. The participants were asked to consider typical and atypical approaches to optimizing carbon, nutrient and energy management at WWTPs while satisfying treatment quality objectives. Typical approaches would include the evaluation of process modeling opportunities and constraints, and incremental resource and carbon management optimization techniques. Atypical approaches were expected to stimulate discussion and development of new concepts. For example, the participants discussed "plant of the future" concepts that can be expected to generate opportunities and research needs related to nutrient and energy sources within treatment plants (whether by heat recovery or energy generation from algae, etc.), changing wastewater characteristics, decentralized treatment (with decentralized or centralized residuals handling) and total water reuse.

The workshop purpose also included the creation of materials that could be compiled into a WERF publication, and that can serve as a guide to WERF as it plans and implements research priorities related to sustainable wastewater treatment. Appendix C for example presents the raw notes and records from the workshop, while the remainder of this publication presents distilled information and written summaries of key topics from the workshop.

CHAPTER 3.0

WHAT IS A TECHNOLOGY ROADMAP?

To prepare for the needs of the future, industry often develops a sector-specific technology roadmap to outline the R&D needs to meet a predicted future. The technology roadmap process has become popular in numerous industrial sectors as they work to reduce their carbon footprint in the carbon-constrained world anticipated in the light of global climate change. Within the water sector, the roadmap process should also consider nutrient recovery, carbon footprint, GHG emissions, and the value of the quality of the water produced. In addition to predicting needs, roadmaps include actionable activities as necessary to achieve the industry vision.

A technology roadmap is also a plan that matches long-term goals with specific technology solutions. The technology roadmap helps reach a consensus about needs and the technologies required to satisfy those needs; provides a mechanism to help forecast technology developments to meet those needs; and provides a framework to help plan and coordinate the required technology developments. This Technology Roadmap Report is for the domestic wastewater reclamation sector but aspects apply to stormwater management, water quality, decentralized treatment and wastewater reuse.

3.1 Specific Objectives of the Technology Roadmap to Optimize Wastewater Treatment in a Carbon and Energy Constrained World

The objectives of this technology roadmap are to:

- Identify long-term sustainability goals for the wastewater sector
- Brainstorm possible technology concepts reasonably expected to produce actionable results that can be implemented
- Consider typical and atypical approaches to optimizing carbon, nutrient, and energy management at WWTPs
- Discuss various "plant of the future" concepts
- Generate opportunities and research needs

3.2 Boundary Conditions

To focus the large number of possible pathways that the roadmap may follow, the working group established some boundaries. Working group participants considered only complete wastewater treatment systems, and focused on the wastewater sector in North America. As an example, these boundaries preclude consideration of sanitation technologies that are not water-or sewer-based, such as composting toilets.

Some other clarifications regarding boundary conditions and related topics to be considered as suggested by the working group:

- Include biosolids transportation and disposal
- Consider both decentralized and centralized treatment
- Consider new technologies that might be needed, but that do not currently exist
- Resource recovery includes water recovery
- Consider wet weather and combined sewer overflows (CSOs) (including potential effluent requirements)
- Consider the collection system – should consider fundamental components including CSOs and sewer management options such as the impact of some odor control techniques on volatile fatty acids (VFA) and carbon constituents of the wastewater
- Consider regional quantity and quality requirements as they may lead to different roadmaps for different conditions
- Include both existing and new technologies
- Initially consider greenfield systems, although the practicality is that existing infrastructure has an existing value that will affect site-specific decisions
- Consider public-private partnerships – especially small plants
- Consider resistance to new technologies by the public or utility management
- Identify barriers related to testing and approval requirements
- Site constraints – greater limitations in future

CHAPTER 4.0

PREDICTED EFFLUENT OR REUSE QUALITY 2040

Participants brainstormed what they felt would be the required treated water quality in 2040. We employed a two-step process, first listing all ideas and opinions, and second grouping the opinions into categories. The second step became necessary when the workshop attendees identified several indirect constraints that would impact the remainder of the workshop activities. The resulting lists therefore encompass not only the treated water quality requirements in 2040 but also other key plant design and operating constraints of the future.

- General Criteria
  - Designated use-based standards
  - TMDL and public health-driven
- Water
  - Microconstituents control
  - Tiered reuse criteria (potable, agriculture, industrial, algae growth)
  - Pathogens – resistance, virulence; targeted pathogens instead of surrogate indicators
  - Nutrients to very low concentrations
  - Low levels of metals
  - Nanoparticle standards
  - Ecologically neutral or enhancing
  - Thermal/energy recovery
- Residuals
  - Tiered reuse criteria (agriculture, industrial products)
  - Microconstituents control
  - Energy recovery
  - Transportation limits
  - Zero residuals – all marketable products
- Air
  - GHG emissions will be limited, only biogenic carbon leaving the site. May be limited on a mass of GHG per capita basis
  - Bio-aerosols transport
  - Odor control, especially at decentralized facilities and collection systems – zero odor detected
- Energy
  - Consumption metrics – kWh/MG or per capita

CHAPTER 5.0

PREDICTED TECHNOLOGIES THAT WILL BE COMMON IN 2040 TO ACHIEVE REQUIRED QUALITY

A three-step process was employed to identify the technologies that will be required and/or common in the future. The first step was to brainstorm ideas and opinions on technologies, to generate an initial master list. The second step involved grouping the technologies into categories, which stimulated further brainstorming within each topic. The third step was to identify those technologies that the attendees felt would require research and development, and to further identify research that would be led by equipment manufacturers and not necessarily be priority research needs for organizations such as WERF. Table 5-1 is the summary of processes that will be best suited to help the wastewater sector become sustainable for energy and resource recovery. Equipment manufacturer-based topics are identified by asterisk.

Table 5-1. List of Key Attributes of Future Wastewater Reclamation Processes.

Treatment Process	***Heat and Energy Recovery***	***Operational***
Use flow equalization	Use solar and wind energy*	Improve operating controls
Enhance primary treatment for maximum carbon capture*	More efficient capture/use of heat and power from residuals	Use smart grid technology at plants
More targeted aeration* and advanced oxidation	Thermal heat recovery*	Recognition of plants with best practices
Greater use of fixed film processes	Hydraulic energy recovery	Greater use of decentralized or satellite facilities under central operations
Superbugs or immobilized enzyme treatment*	Energy storage	***Other***
Extract embedded energy in wastewater (i.e., microbial fuel cells)*	New carbon sources (i.e., waste from food industry or food grinders)	Residual carbon sequestration
Greater use of media filtration and membranes*	Reduce energy requirements for treatment	Collect/use/treat sewer off-gas
Greater use of anaerobic treatment	***Nutrient Recovery***	
Sidestream treatment [such as anaerobic ammonium oxidation (anammox)]	Recovery of N, P, and S	
Use nanomaterials for treatment	Autotrophic and phototrophic nutrient removal (i.e., more use of algae)	

The following listings provide more detail on the specific areas that the expert working group felt must occur to meet sustainable goals. The technologies that are expected to be developed by equipment manufactures are identified in the lists below by an asterisk.

Utility Management and Watershed-Based Topics

- Produce toolbox of technologies and methods that recognize regional differences caused by wet weather, reuse demands, etc. It is recognized that we cannot identify a standardized plant or technology to be used for all applications.
- Watershed/regional residuals co-management
- Decentralized treatment with centralized residuals
- Paradigm – water extraction versus treatment
- Minimize infiltration and inflow (I/I) in collection systems
- Public education becomes a major focus
- Water conservation program effects
- Green energy sources*
- Carbon dioxide (CO_2) cycled to algae contactor, algae to biofuels
- Co-processing of food wastes
- Thermal heat recovery*
- Collection system heat recovery*

Wastewater Treatment Topics

- Membranes*
- Better aeration*
- Sidestream treatment
- Anaerobic treatment, including sulfur and methane recovery
- Friendlier disinfection (cheaper, non-chlorine-based)*
- Superbugs/immobilized enzyme treatment*
- Natural systems
- Photosynthesis
- Fixed film
- Advanced oxidation
- Off-gas collection and reuse
- Media filtration (other than membranes)*
- Physical/chemical treatment*
- Adsorption*

Biosolids and Residuals Topics

- Subsurface residuals injection
- Cogeneration*
- Thermal biosolids processes*
- Mining of inorganics – N, P, S
- N and P recovery from centrates
- Photosynthetic treatment of centrate
- Biogas treatment*

CHAPTER 6.0

RESEARCH AND BREAKTHROUGHS NEEDED TO MAKE FUTURE WASTEWATER TREATMENT PLANTS SUSTAINABLE

Following a general discussion of the need to combine the technologies into full-plant systems, the working group brainstormed a broad list of research needs, decision-making methods, and quantification tools that will be required to allow utilities and operators to achieve the nutrient- and energy-neutral "plant of the future" while achieving water quality objectives. A ranking exercise then identified the highest ranked needs (for which there was a consensus), as well as a second tier (that were supported by a large majority of working group participants). During the discussions, a consensus was also reached that communications and public education will play a large role toward the implementation of "plant of the future" concepts. Although included below, expanded communications and public education discussions are included in a separate section of this report.

The working group used a breakout group approach to identify prioritized lists of technology developments and technology research that is necessary to achieve energy and carbon neutral plants of the future. Each breakout group was asked to consider three priority categories: barriers that must be broken through research to develop new technologies; incremental developments to current technologies that must occur to improve efficiencies; and whole-plant concepts that optimize the complete plant rather than optimize individual components of an assumed whole-plant system. These breakout sessions provided three products: lists of prioritized technology research needs, preliminary scopes and budgets for the prioritized technology research topics, and preliminary concepts for complete plant process flow diagrams.

The preliminary scopes and budgets for the highest-prioritized technology research topics are included as Appendix A to this report. Preliminary concepts for complete plant process flow diagrams for various plants of the future are included in Chapter 8.0, with the full submissions included as Appendix B.

Figures 6-1 to 6-4 present the prioritized technology developments and research in graphical form. Each figure addresses a specific grouping: treatment processes, equipment, management, and other activities.

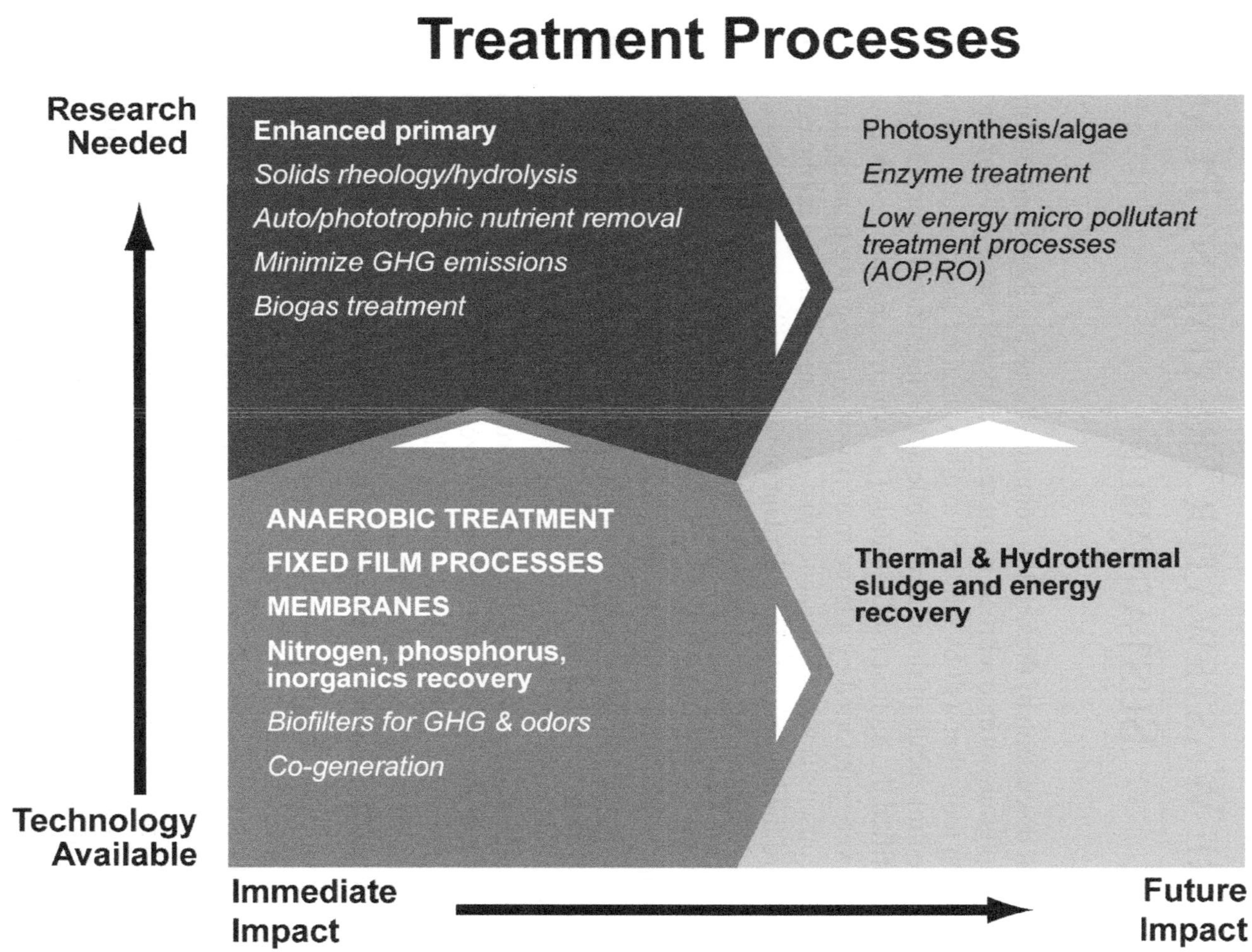

KEY:
BOLD CAPITALS - Identified by all 4 work groups **Bold Lower Case** – Identified by 3 of 4 groups Lower Case – 2 of 4 groups *Italicized* – 1 group

Figure 6-1. Treatment Process Prioritization.

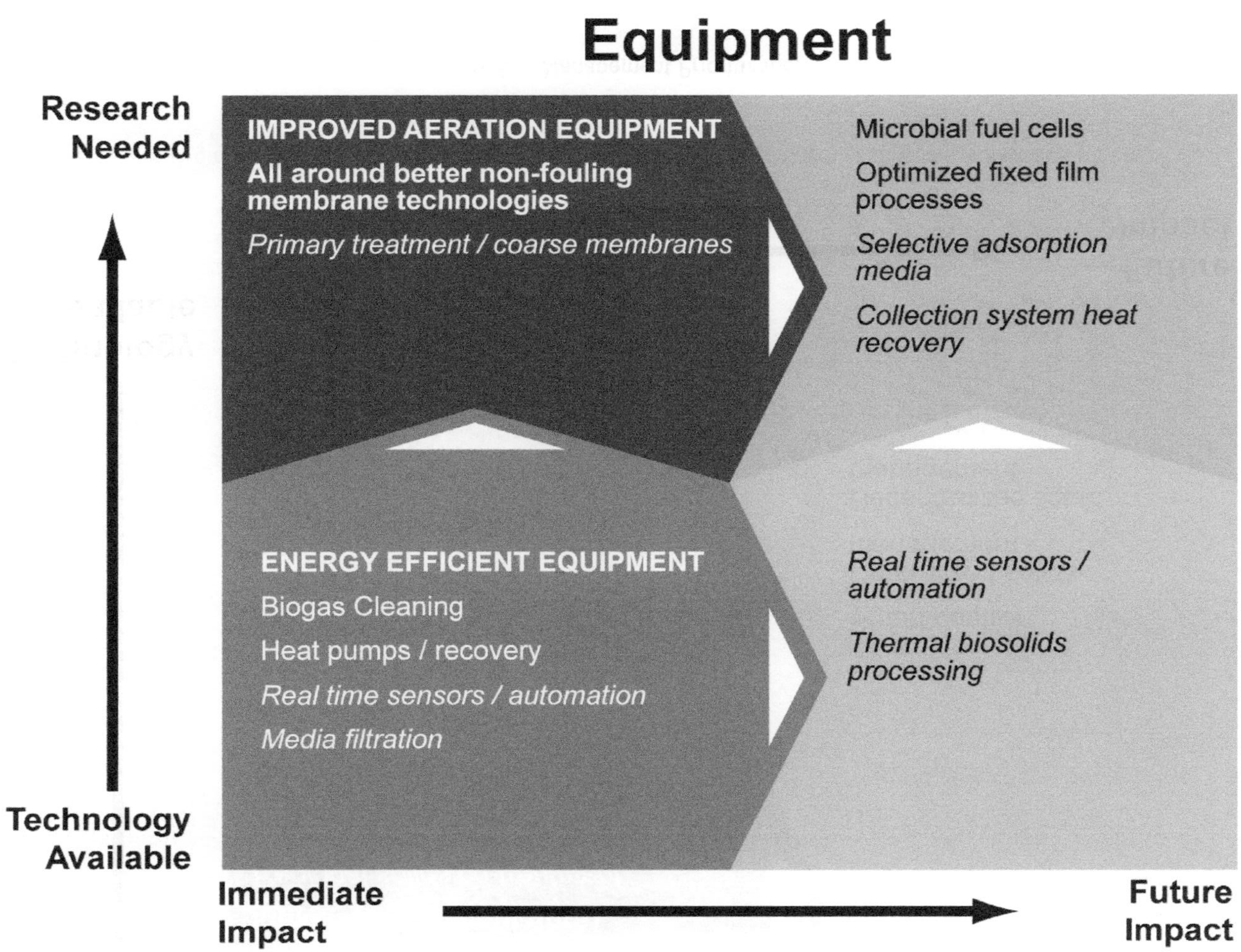

KEY:
BOLD CAPITALS - Identified by all 4 work groups **Bold Lower Case** – Identified by 3 of 4 groups Lower Case – 2 of 4 groups *Italicized* – 1 group

Figure 6-2. Equipment Prioritization.

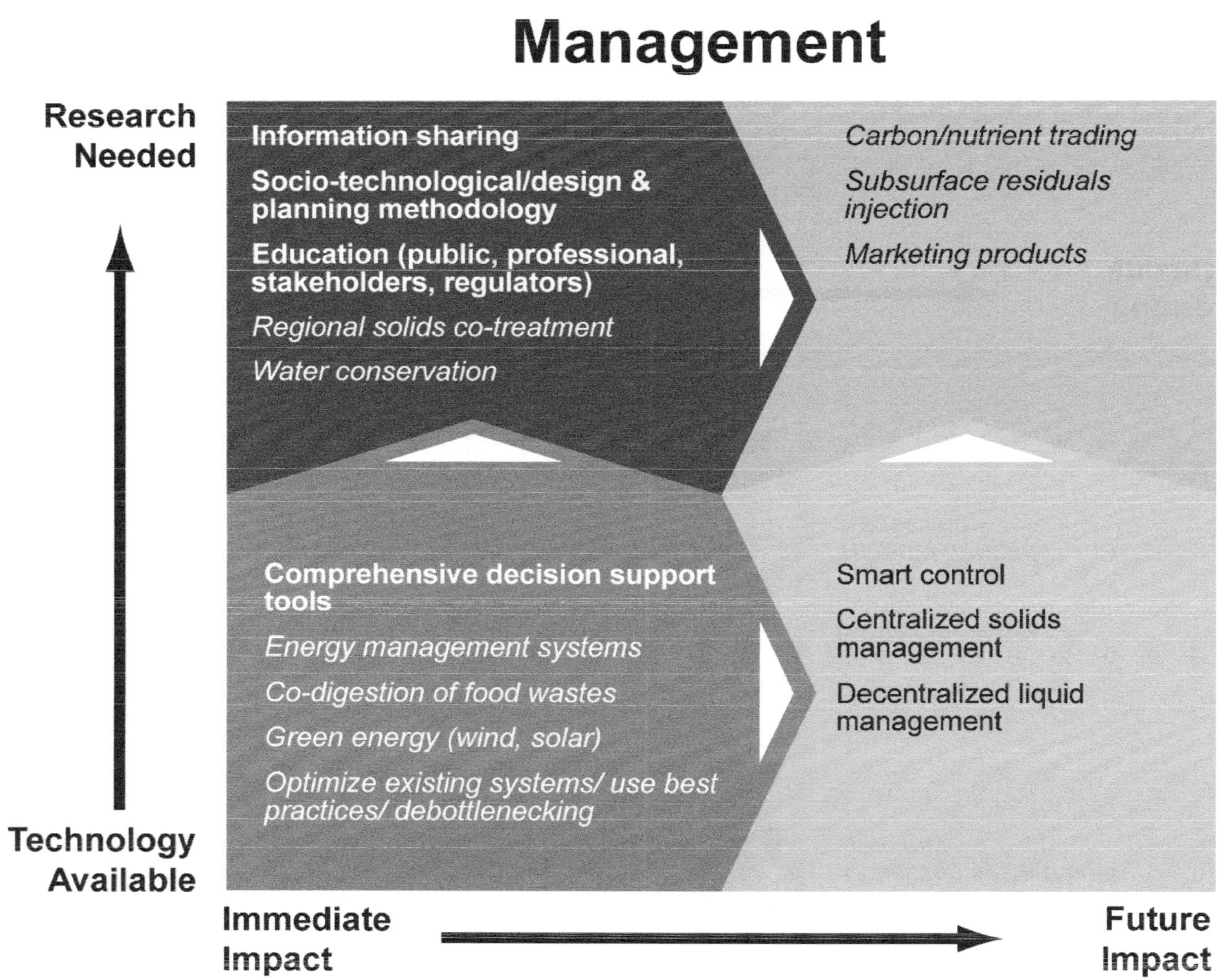

KEY:
BOLD CAPITALS - Identified by all 4 work groups **Bold Lower Case** – Identified by 3 of 4 groups Lower Case – 2 of 4 groups *Italicized* – 1 group

Figure 6-3. Management Prioritization.

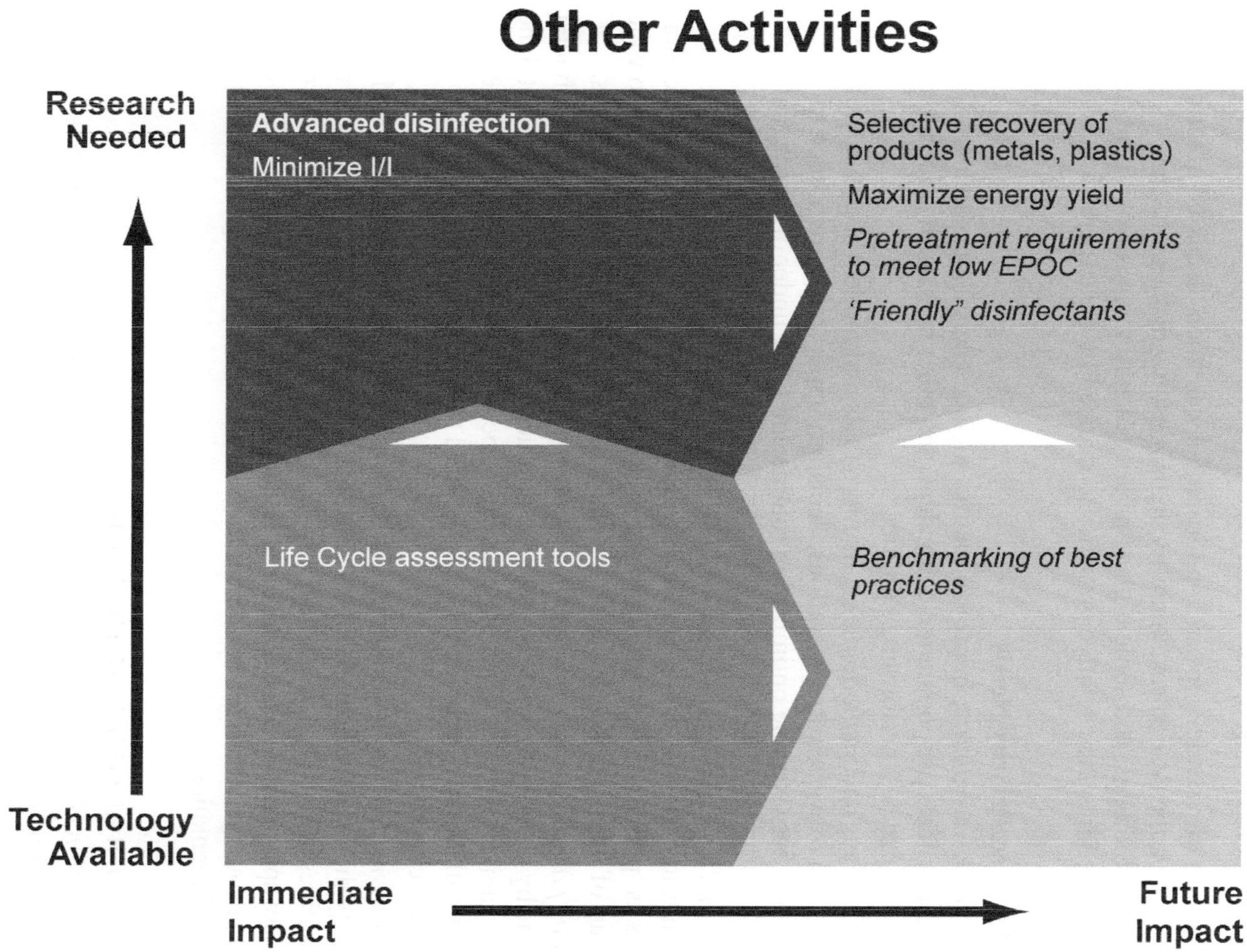

KEY:
BOLD CAPITALS - Identified by all 4 work groups **Bold Lower Case** – Identified by 3 of 4 groups Lower Case – 2 of 4 groups *Italicized* – 1 group

Figure 6-4. Other Activities Prioritization.

Each figure identifies the prioritized research within four quadrants, reflecting four combinations of technology status and timing. The vertical axis represents the technology status, differentiating between technologies that are already proven and available, and those for which significant research is needed. The horizontal axis represents the timing of the return on implementation of the technology, differentiating between technologies that will have immediate versus future impact. Ideas that came up in all four groups are bold and capitalized, in three groups are bold lower case, in two groups are lower case and in one group are italicized.

The identified lists of research gaps and opportunities were then discussed in terms of their expectation to achieve the desired sustainability goals. Ensuing discussions then consolidated the experts' thoughts, and some highest priority overall themes were identified. The prioritized themes, grouped into their relative priorities as decided by the experts, are as follows:

Highest Ranked Identified Priority Needs

- *Improved metrics for sustainability.* Improved metrics include energy measurements (such as kWh/MG), CO_2 equivalents per population equivalent, and other metrics that need to be standardized to recognize best performing facilities.
- *Decision support systems and methodologies.* As more tools and models become available to guide the optimization of wastewater operations, decision support tools are needed to navigate the tradeoffs between one optimization routine and other under different goals and criteria.
- *Public education and rebranding of our sector.* The general public does not appreciate the value of the clean water and the service provided by this sector. Instead there is a perception of all dischargers as 'polluters.' Rebranding of our sector from wastewater treatment to water reclamation is part of future public education to ensure public support to these services.
- *New R&D funding paradigm – new product development.* Research to date has been divided between that done by industry to create marketable products and technologies, and that done by utilities and organizations such as WERF to create processes and technologies made available to all. A collaborative approach is now required to accelerate research and investment in new technologies.
- *Risk management is needed, versus risk aversion.* Public utilities are required, first and foremost, to satisfy their discharge permit requirements. This has created a mindset of risk aversion; with utilities reluctant to implement new technologies if permit compliance might be affected. A new approach to risk management is required, including both the utility and the regulator, to encourage the use of new technologies, particularly during the initial phases of technology implementation.

Other Strongly Supported Identified Priority Needs

- *Further reliance on biotechnology.* The sector should not lose sight of the fact that biotechnology developments have benefited the water industry throughout its history, perhaps more than any other knowledge area.
- *Research/knowledge cross-over.* Significant applicable research and knowledge already exists outside the domestic water sector. Technologies and research from the industrial and energy sectors, as examples, should be accessed.

- *Feedstock to plant (additional carbon sources).* The addition of high carbon wastes such as greases and fat to anaerobic digesters will improve energy balances. The use of household waste grinders could be re-evaluated.
- *Improved technology transfer.* Knowledge and experience from outside North America should be accessed. Research coordination among the various North American entities should be encouraged.
- *Revised regulatory approach – incentive-based rather than punitive.* The current quality-compliance only approach could be augmented to include rewards or offsets for achieving various metrics such as energy use per capita, nutrient recovery or similar parameters.

The Broad List of Identified Needs to Achieve the "Plant of the Future"

- Flow equalization to allow more optimized plant operations
- More use of co-generation to produce combined heat and electrical power
- Improved primary treatment to increase carbon diversion to solids processing
- Maximize carbon capture in primary clarifiers and elsewhere delivered to anaerobic digestion to maximize biogas production
- Re-visit food grinders to increase carbon sources within wastewater
- Reduce energy required for treatment through process and equipment developments
- Extract embedded energy including heat recovery from liquid streams
- Green energy (solar, wind)
- Transportation outside the fence should be considered when planning biosolids and residuals systems
- Anaerobic treatment of wastewater to convert soluble BOD and chemical oxygen demand (COD) to biogas
- Microbial fuel cells
- Chemical enhanced primary treatment, and better chemicals to increase carbon diversion to solids processing
- Improved operating controls
- Centralized operation of decentralized facilities
- Improve energy-efficiency of satellite facilities when treating minimal flows to meet local reuse demands
- Operator education
- Thermal heat recovery such as direct heat exchange with plant effluent
- Improve efficiency at low plant flows
- Public outreach/education – recognize the value of reuse water and recovered products
- Water quality-driven regulations to satisfy or enhance the intended use
- Energy storage
- Hydraulic energy recovery such as effluent turbines
- Nanomaterials for treatment
- Geological residuals carbon sequestration
- Carbon/nutrient trading/regional management
- Integrate with natural systems
- Sewer off-gas treatment
- Better inorganics removal from mixed liquor suspended solids
- Wet air oxidation, supercritical oxidation

- Comprehensive decision support tools to balance carbon/water/energy
- Smart grid technology within plants
- Solids hydrolysis
- Improved biosolids water removal efficiency
- New carbon sources for energy recovery such as fats and greases
- Improved efficiency of power generation from biogas and biosolids
- Identification and recognition of existing plants having best practices
- Systemized instrumentation and controls
- Autotrophic and phototrophic nutrient removal
- Multiple barrier disinfection
- Define inclusive boundary conditions – is anything really an externality?

CHAPTER 7.0

COMMUNICATIONS AND PUBLIC EDUCATION

During the Technology Roadmap discussions, the working group reached a consensus that communications and public education will play a large role toward the implementation of Technology Roadmap. The working group agreed that the Technology Roadmap needs to encompass these aspects:

Develop design and planning methodologies that optimize both social and technological objectives

- Consider all potential contributors and users of the energy, water, and nutrients received and recovered at the plant
- Stakeholder inclusion and engagement
- Watershed approach including industrial activities, current and future users
- Iterative process
- Get it working in the field (highly interdisciplinary)

Education

- Public
- Public education (pop culture, wiki, etc.)
- Future water professionals
- Regulators, policymakers
- Other stakeholders
- Sustainability principles and involvement – Possibly designate a coordinator
- Marketing of products recovered during treatment
- Means for trusted information sharing, communication and technology evaluation with the public

CHAPTER 8.0

"PLANT OF THE FUTURE" CONCEPTS

A variety of plant-wide concepts were submitted by some attendees following the workshop. The submissions as received are included as Appendix B.

The submitted 'plant of the future" concepts shared several similarities, with creative approaches identified to achieve the overall functional objectives. Those objectives included the achievement of reuse-quality water by an energy self-sufficient plant with the greatest possible recovery of nutrients. The minimization of carbon footprint and GHG emissions was implicitly assumed by each, possibly reflecting the preliminary state of research in these two particular areas.

The similarities among the submissions generally included an overall vision of the complete plant and its energy and nutrient balance; physical separation of particulate carbon from the raw wastewater; two-stage biological treatment to remove carbon (BOD) initially before nutrient removal; provisions for advanced treatment to meet future requirements; conversion of particulate carbon to biogas or biodiesel fuels; separate treatment of sidestream recycle flows from residuals processing before re-combination with the liquid treatment train; and beneficial use of any remaining residuals. Other components were mentioned, such as thermal heat recovery from the wastewater liquid stream, improved disinfection technologies, green energy production, and improved staff training, the details of which can be found in Appendix B.

Creating an overall vision of the complete plant fostered creativity, and in some cases identified desirable technologies for which fundamental research is still required. Submissions developed differing emphasis, such as a reliance on membrane technologies instead of biological treatment, or a focused objective to maximize biogas production.

Physical separation of particulate carbon from as the raw wastewater, as an initial treatment step, will increase the mass of raw carbon diverted into the plant residuals processing area, with the result being increased energy recovery potential such as through the production of biogas. Physical separation can be by traditional primary clarification, or by other potentially more efficient separation technologies for which additional research and technology development is required: chemically-enhanced clarification, direct granular media filtration, membrane separation and/or combinations of each. One submission assumed that membrane separation technology will advance to the point where biological treatment will not be required, resulting in a physical separation-based liquid treatment concept.

Two-stage biological treatment addresses carbon and BOD removal before nutrient reduction. The initial stage of carbon and BOD removal typically adopted one of two options: anaerobic treatment for the direct conversion of BOD to biogas, and low solids retention time (SRT) treatment to minimize the aerobic destruction of carbon and BOD within the aeration tanks so that the captured carbon remains available in the waste activated sludge for conversion to biogas within the residuals processing area of the plant. The second stage of biological treatment, to reduce and/or recover nutrients, was identified as being dependent on the quality of the reclaimed or reuse water desired and/or the value placed on the nutrients recovered. Some submissions identified multiple pathways depending on the treatment objective.

Most submissions identified provisions for advanced treatment to meet future requirements. Advanced oxidation, improved disinfection technologies, and reverse osmosis separation were most common, with an implicit assumption that these would become required to reduce microconstituents and total dissolved solids concentrations.

Within the residuals processing area, the conversion and use of biogas to create combined heat and power was a popular choice. The inclusion of algae to biodiesel fuel concept, and the extraction of nutrients from sludge, was also mentioned. The inclusion of additional wastes having high potential fuel value, such as waste greases, food wastes or septage, was identified as a means to increase fuel production through residuals processing.

The separate treatment of sidestream recycle flows from residuals processing before re-combination with the liquid treatment train was identified by consensus of the experts at the workshop, however only preliminary concepts were included within the submitted concepts. The use of a nitritation-denitritation process such as employed at the Strass, Austria, plant was favored.

The beneficial reuse of the biosolids produced was assumed in most submissions. The alternative combustion of the final residuals with recovery of heat energy was also mentioned.

8.1 One Team's Vision of the Future for WWTPs

This team's vision, illustrated in Figure 8-1, maximizes biogas production from carbon by including anaerobic treatment steps within both the liquid treatment stream and the sludge stream. Both membrane and advanced oxidation steps are included, suggesting this option is suitable for the production of very high quality water. This concept is perhaps best suited to warm wastewaters without high I/I because of the anaerobic liquid treatment step, but has the potential to generate additional biogas that can be captured for energy recovery while the anaerobic conditions eliminate the costly and energy intensive requirements for oxygen transfer.

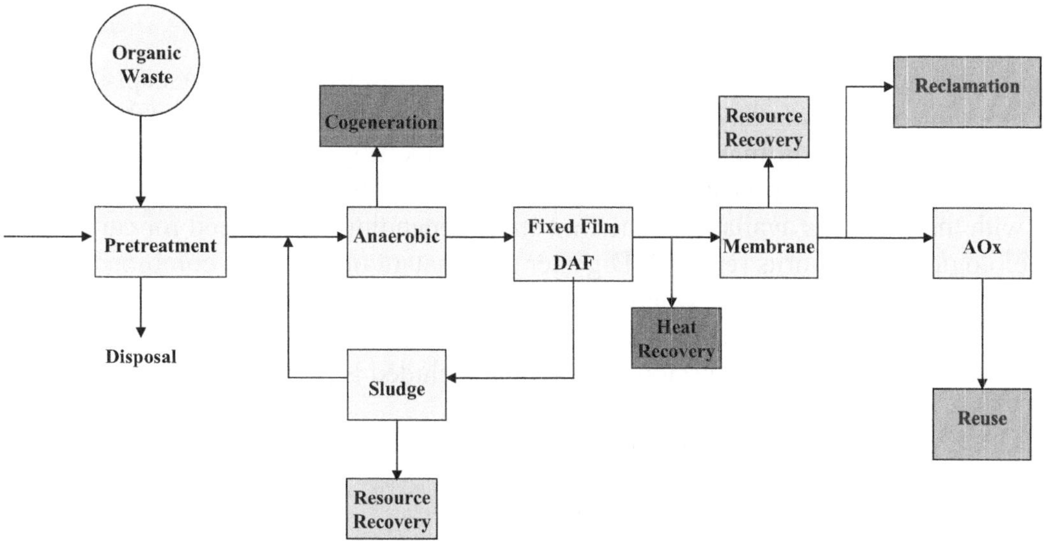

Figure 8-1. One Team's Vision.

8.2 A Vision of the Future by Joe Zuback

The Joe Zuback concept shown in Figure 8-2 is a filtration and membrane-based liquid treatment approach that requires research to overcome membrane maintenance and flux issues. The "plant of the future" illustrated below separates particulate without biological oxidation, and directs that particulate to anaerobic digester and biodiesel technologies to maximize the conversion of all plant particulate carbon to biogas and to biodiesel.

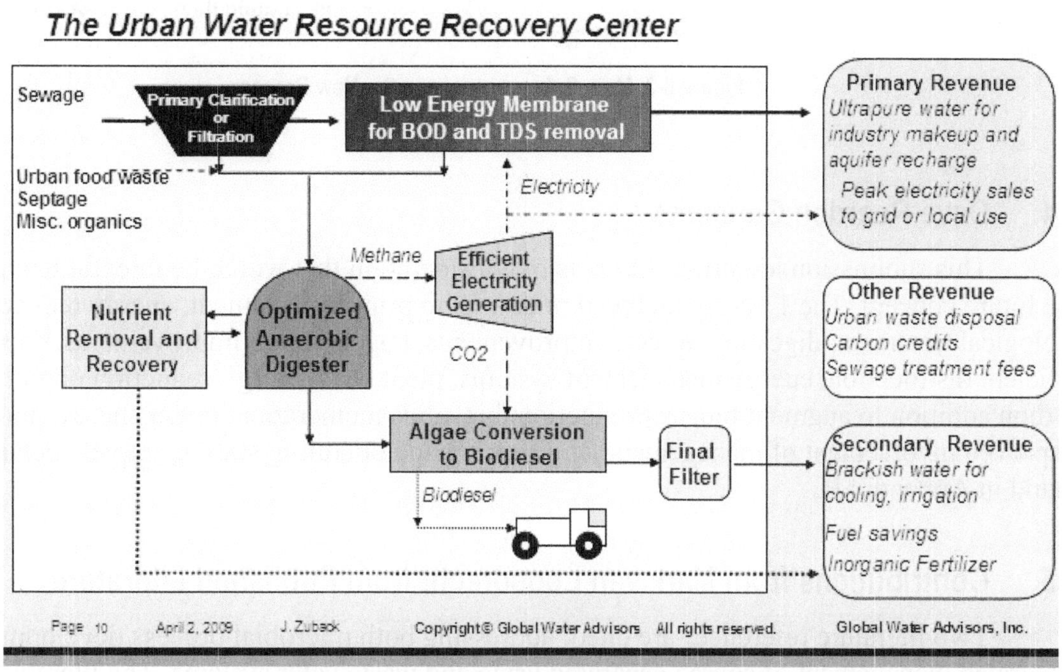

Figure 8-2. Joe Zuback Submission.

8.3 "Plant of the Future" by Cao Yeshi

As also done in the Joe Zuback concept, the "plant of the future" concept illustrated in Figure 8-3 reduces the plant oxygenation energy requirements by maximizing the removal of organic particulate in the primary treatment units and diverting that material to anaerobic digesters for biogas production and energy co-generation. The liquid train is shown in two stages, with three options available in the first stage depending on the need for carbon, nitrogen and/or biological phosphorus removal. Digester supernatant or filtrate is combined with the liquid stream in the second stage, where aerobic heterotrophic organic transformation is minimized and the oxygen needed for nitrogen removal is reduced by using a nitritation process such as anammox. The full submission with details included is presened in Appendix B.

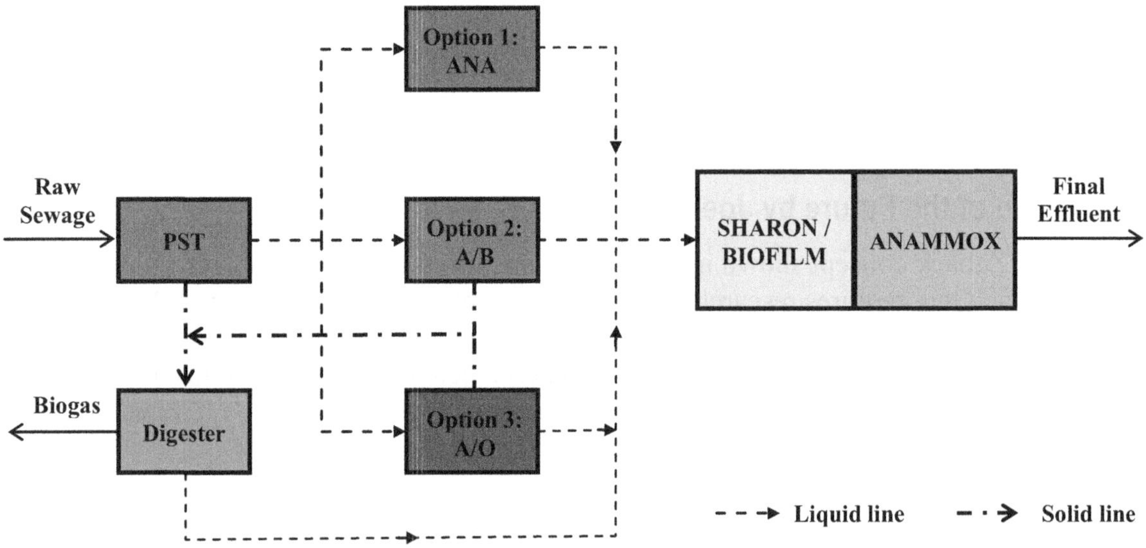

Figure 8-3. Main Sub-processes in the New Process.

8.4 Dave Reardon Concepts

This submission identified 11 areas of consideration that would be integral to a plant of the future concept. The 11 areas included maximizing primary treatment, energy reduction in biological treatment, digestion process improvements, cogeneration improvements, energy-efficient disinfection, energy management systems, plant wide energy-reduction opportunities, carbon addition to augment biogas production, biosolids incineration, green energy, and improved involvement of management, engineering and operating staff. Complete details can be found in Appendix B.

8.5 Contributions from Mark van Loosdrecht from Published Literature

Two literature references are cited, addressing both microbial process developments and integrated management of source-separated urine and wastewater systems. The literature sources are included in Appendix B.

CHAPTER 9.0

NEXT STEPS

As closure to the workshop, attendees were advised of the next steps planned for the full roadmap development.

1. Gather suggested complete system "plant of the future" concepts from attendees following the workshop. Attendees were encouraged to record any plant flow diagram concepts developed during the final breakout sessions, and to forward them to the workshop facilitators. The intent was to capture process flow diagrams for complete plant, and to integrate compatible concepts within a complete plant concept, with the objective of creating energy self-sufficiency. Although the submissions have been included in this report, it is recognized that the submitted "plant of the future" concepts still require peer review, as well as the verification of complete plant mass balances and performance assumptions. That quality control and peer review is beyond the scope of this report, and rather the raw concepts are included in this workshop documentation to stimulate discussion and thought.

2. WERF will review the draft planning documents for each prioritized research need, including scope, objectives, deliverables, and budget estimate. These could then become proposals for future research.

3. Using the prioritized lists from the workshop as a guideline and framework for decision, WERF will create a roadmap of research and development (R&D) activities to achieve resource and energy self-sufficiency in the plants of the future while at the same time minimizing carbon footprints and GHG emissions.

APPENDIX A

PRELIMINARY BUDGETS AND SCOPES FOR PRIORITIZED RESEARCH TOPICS

A.1 Nutrient Recovery (N, P, S)

General Scope/Objectives

- Economically recover N, P, S
- Reduce Nutrients from biosolids, gas, recycle streams & wastewater
- Produce a usable product
- Develop markets for products

Benefits/Outcomes

- Reduce treatment costs
- Increased process and /or plant capacity
- Revenue
- Recovery of a non-renewable resource

Research Sponsors

(Utilities, Federal Government, Academia, Vendors, Professional Organizations)

P – Primary (one only), S – Secondary (unlimited)

P – Academia Utility collaboration

S – Vendors

S – Federal Government

Sequencing
Preceding Activities

- Basic Research
- Peer Reviewed & Grey Literature search(European, Japan,)

Subsequent Activities

- Pilot studies
- Full Scale implementation

Duration___5___years

Level of WERF Involvement ____20____%

A.2 Optimization of Reduced Carbon to Energy (Liquid Phase)

General Scope/Objectives

- Maximize solids removal in primary treatment – with anaerobic digestion
- Maximize the use of liquid phase anaerobic treatment for carbon use
- Effective conversion of fats, oils, and grease (FOG) to energy – liquid phase separation and either digestion for methane production or liquid fuel production
- If denitrification is required, use reduced sulfur formed in anaerobic treatment as an electron donor. Use aerobic treatment for nitrification only. More difficult to integrate with biological phosphorus removal.
- Manage/collect sources of fugitive methane emissions (e.g., collection system, aeration tanks, clarifiers, channels, etc)
- Water conservation, source/urine separation, etc with the objective of concentrating carbon in waste, while eliminating other constituents (e.g., N&P), thus improving the C:N and C:P balance.

Benefits/Outcomes

- Promotes energy and carbon neutrality
- Minimizes aeration demand
- Maximizes methane production
- Minimize sludge production

Research Sponsors

(Utilities, Federal Government, Academia, Vendors, Professional Organizations)

P – Primary (one only), S – Secondary (unlimited)

P – Federal

S – Utilities

S – WERF

Sequencing
Preceding Activities

- Develop cost-effective enhanced primary treatment
- Improve understanding and development of low temperature anaerobic treatment processes and integration within existing process trains
- Integration of anaerobic treatment with denitrification over reduced sulfur and remaining methane (CH_4) in solution
- Understand value of carbon recovered in primary treatment versus carbon use for denite and bio-P
- Integration of low temperature anaerobic treatment processes with existing biological treatment infrastructure
- FOG to energy – separation from waste, transfer to digestion (maybe not a research issue – technology transfer and education)

Subsequent Activities

- Small plant demonstration project

- Collaboration with South American and other warm climate utilities that have good experience with anaerobic treatment
- Collaboration with industrial users of anaerobic liquid treatment processes (e.g., upflow anaerobic sludge blanket (UASB), fixed film anaerobic, hybrid suspended/fixed anaerobic)

Duration ___10___Years

Level of WERF Involvement ___50___%

A.3 Heat Recovery

General Scope/Objectives

Wastewater contains recoverable heat compared ambient water sources, so estimates are that this recoverable heat significantly greater than the energy recoverable from carbon. This concept has implemented already in the US once in King County, WA and well in Europe.

- Develop technologies for Efficient Recovery of Thermal Energy from Wasterwater and Side-streams
- Develop users for the low grade heat recovered
- Understand the energy and cost models to make this concept economically feasible
- Develop a life cycle assessment (LCA) model implementation

Benefits/Outcomes

- Cheap heat source for neighbors and plant
- Integration of treatment plant with community
- Carbon neutral or positive
- Energy independence

Research Sponsors

(Utilities, Federal Government, Academia, Vendors, Professional Organizations)

P – Primary (one only), S – Secondary (unlimited)

P – Utilities, large and small

S – Federal Government (Department of Energy),

S – Thermal Equipment Manufacture

S – Consulting Engineer

Sequencing
Preceding Activities

- Understand the energy and cost models to make this concept economically feasible
- Develop a LCA model implementation
- Develop technologies for Efficient Recovery of Thermal Energy from Wasterwater and Side-streams
- Test new and validate the technology
- Develop users for the low grade heat recovered

Subsequent Activities

- Report and communication to subscribers on the success and issues on the project.

Duration ___3___ years

Level of WERF Involvement ___20___%

A.4 Optimization of Carbon to Energy (Solids) Thermal Processes

General Scope/Objectives

- Evaluation new developments in thermal technologies for biomass conversion and their applicability to sludge from WWTP. For example advanced pyrolysis process with bio-char formation, flash pyrolysis for biofuels, subcritical-water drying, pressure gasification, wet air oxidation, supercritical oxidation, co-firing
- Evaluation of thermal technologies for co-processing of sludge and biomass
- Evaluate destruction of emerging contaminants present in the sludge by advanced thermal technologies
- Evaluate potential for recovery of products, Phosphorous, Metals, bio-char as part of the process
- Evaluation of energy, carbon and water footprint of existing thermal technologies for sludge or sludge-biomass co-processing
- Demonstration of selected thermal technologies for sludge processing in WWTP

Benefits/Outcomes

- Provide cost-effective alternatives for sludge processing that minimize carbon, energy and environmental impacts
- Provide alternatives to the industry for dealing with microconstituents in the sludge
- Enhance the potential for product –resource recovery from WWTP: Energy, P, Metals

Research Sponsors

(Utilities, Federal Government, Academia, Vendors, Professional Organizations)

P – Primary (one only), S – Secondary (unlimited)

P – WERF

S – Vendor in some cases

S – Utility support,

S – Federal EPA

S – Academia, in kind support

Sequencing
Preceding Activities

- Agreement on protocols for C, Energy, Water footprint evaluation (this can be proceeding in parallel)

Subsequent Activities

- Support on Evaluation of Demonstration Units Performance.

- Diffusion of Evaluation Results

Duration __10__ Years

Level of WERF Involvement ____Est 60____%

A.5 Optimization of Carbon to Energy/Anaerobic Digestion

General Scope/Objectives

- Objective
  - Increase conversion efficiency of carbon to methane in anaerobic digesters processing waste solids and codigest feed stocks
- Scope
  - Evaluate sludge conditioning methods to hydrolyze biomass to increase availability of carbon for anaerobic conversion and select preferred method based on effectiveness, cost, and energy requirements. Test in full scale or pilot scale digesters.
  - Evaluate and methods to prethicken solids feed or to provide recuperative thickening to increase SRT of existing digesters and select process for testing based on effectiveness, cost, and energy requirements. Include method #1 and test in full scale or pilot scale digesters.
  - Evaluate and develop biotechnology methods to enhance hydrolysis of lignin and cellulose compounds by enzyme or bioaugmentation to increase carbon conversion efficiency. Test in full scale or pilot scale digesters.

Benefits/Outcomes

- Increase methane production to reduce wastewater plant external energy demands
- Provide biomethane fuel for city/public vehicles
- Net reduction of GHG and carbon footprint by using waste sources for energy and use of less fossil fuels

Research Sponsors

(Utilities, Federal Government, Academia, Vendors, Professional Organizations)

P – Primary (one only), S – Secondary (unlimited)

P – U.S EPA

S – WERF

S – Utilities

Sequencing
Activities

- Evaluate solids conditioning and pre- or recuperative thickening methods and select preferred process(es)
- Test and demonstrate enhanced digestion methods in full scale or pilot scale digesters (may test 1 or more processes)

- Carry out fundamental biotechnology research to develop possible commercial methods to enhance lignin/cellulose conversion.
- Test and demonstrate enhance digestion by enzyme or bioaugmentation in pilot scale digesters.

Duration

Activity 1 and 2 – 3 years

Activity 3 and 4 – 4 years

Level of WERF Involvement ____33 to 50___%

Total budget will be in range of $1.0 to 2.0 million. Note that research may occur independently of WERF, requiring only a verification role for WERF.

A.6 Microconstituents Program

General Scope/Objectives

- Objective: develop/validate methods to meet future stringent requirements for removal of microconstituents while meeting energy/carbon goals and other constraints (ease of implementation, cost)
- Technologies may include advanced oxidation, biofilms, reverse osmosis, adsorption, and others (forward osmosis, nanofiltration)
- Low energy, carbon neutral
- May include pretreatment techniques to improve removal process
- Subprogram: removal of non-organic microconstituents (metals, nanoparticles)

Benefits/Outcomes

- Protection of public health and environment
- Ability to meet future water quality requirements
- Effective, efficient, and robust methods
- Minimal impact (energy, carbon, cost, social)

Research Sponsors

(Utilities, Federal Government, Academia, Vendors, Professional Organizations)

P – Primary (one only), S – Secondary (unlimited)

P – WERF – coordinate effort, facilitate agreement of stakeholders, some funding

S – Federal EPA

S – Other professional organizations – WaterReuse Foundation, WRF, IWA, Global Water Research Coalition

S – Vendors [ultraviolet (UV), membrane, adsorbents, etc.] – technology expertise and funding

S – Academia – base technologies, cooperative R&D and in-kind support

S – Utilities

S – State governments

Sequencing
Preceding Activities

- "Indicator compounds," categories of compounds, other metrics – stakeholder agreement, standard methods of analysis (including bioassays)
- Decision criteria (success criteria for the process, energy/carbon metrics, to feed into decision tools later)
- Assessment of current state of the art and identify gaps
- Lab research for screening and development of solutions to fill gaps

Subsequent Activities

- Field testing of prototype systems
- Development and agreement on design/validation protocols
- Full scale implementation and case studies
- Communication, information sharing, facilitate implementation

Duration

5 yrs to field testing + 5 years to case studies of accepted systems years

Level of WERF Involvement __33__%

A.7 Aeration

General Scope/Objectives

- Minimize energy input/maximize transfer (kWH/kg O_2)
- Scope covers compression, deliver, controls – soup to nuts, from blower (including variable frequency drives) to controls, sensors and instrumentation
  - Special attention to sensors – non-fouling
- Different types of equipment – jets, mixers, etc.
- Figure out what types of equipment work in parts of your tank with low Alphas (hybrid aeration systems). How do you manipulate alpha?
- Investigate balance between air for aeration and air for mixing or use more efficient means for mixing

Benefits/Outcomes

- Reduced power consumption
- Better treatment (potentially)
- Lower operations and maintenance (O&M) because of reduced labor (more automation)
- Longer duty life of equipment

Research Sponsors

(Utilities, Federal Government, Academia, Vendors, Professional Organizations)

P – Primary (one only), S – Secondary (unlimited)

P – WW Utility (w/ Academia)

S – Vendor

S – Fed governments (Department of Energy?)

S – Power utilities and/or Electric Power Research Institute, state energy offices New York State Energy Research and Development Agency

Sequencing
Preceding Activities

- Current usage (benchmarking of utilities' aeration systems/equipment)
- Theoretical limit of oxygen transfer

Subsequent Activities

- Manipulation of alpha
- Put in a hybrid system that takes advantage of alphas.
- Demonstration of optimum control methods (need a simple, reliable system)

Duration __2 to 3__ years – with interim milestones

Level of WERF Involvement ___TBD___%

WERF coordinates and leads. Note that research may occur independently of WERF, requiring only a verification role for WERF.

A.8 Fixed Film

General Scope/Objectives

- Increase existing infrastructure capacity
- Provide guidance for retrofit vs. new construction
- Improve/develop standalone fixed film technologies
- Secondary, tertiary, sidestream
- Media improvements

Benefits/Outcomes

- More consistent performance (through peaks, for example)
- More compact treatment footprint
- Increase reliability
- Lower sludge production
- Lower energy use

Research Sponsors

(Utilities, Federal Government, Academia, Vendors, Professional Organizations)

P – Primary (one only), S – Secondary (unlimited)

P – Vendors/equipment suppliers

S – Utilities

S – Federal government

Sequencing

Preceding Activities

- ♦ Summarize existing technologies – what's working, what's not working/status of technology
- ♦ Identify known information regarding impact of process configurations on the fundamental processes
- ♦ Visit existing facilities that have "real" operators running the plant (not vendors)

Subsequent Activities

- ♦ Media development
- ♦ Process configurations

Duration __3-4__ years

Level of WERF Involvement __50__ %

A.9 Disinfection

General Scope/Objectives

- ♦ Determine alternative disinfection methods (e.g., radio frequency, electric pulse, e-beam)
- ♦ Decrease energy consumption
- ♦ Decrease byproduct formation
- ♦ Better understanding of pathogen destruction (vs. inactivation), survival and development of resistance
- ♦ Reduce infrastructure (footprint) required for treatment
- ♦ If we use membranes, do we approach disinfection separately to address viruses only?

Benefits/Outcomes

- ♦ Energy and material (chlorine) conservation
- ♦ Reduced byproduct formation
- ♦ Reduced footprint/capital infrastructure/cost
- ♦ Reduced public health risk

Research Sponsors

(Utilities, Federal Government, Academia, Vendors, Professional Organizations)

P – Primary (one only), S – Secondary (unlimited)

P – Utilities, Federal Government

S – equipment vendors

S – chemical suppliers

S – Academia

Sequencing
Preceding Activities

- Look at new technologies – mechanisms to kill microorganisms
- Shortlist mechanisms of destruction and resistance

Subsequent Activities

- Basic research – lab scale to pilot to full scale
- Study destruction methods (vs. inactivation)
- Identify specific emerging pathogens
- Differentiate between influent driven and plant derived pathogens, resistance determinants

Duration ___4 to 6___ years

Level of WERF Involvement ___50___%

A.10 Membranes

General Scope/Objectives

- Reduce LCC (energy, fouling mitigation, membrane life, system cost)
- Achieve anticipated future effluent requirements
- Maximize soluble carbon (in sewage) capture for energy production (methanization, microbial fuel cells, etc.)
- Assurance of public health protection for systems without biological treatment of liquid stream

Benefits/Outcomes

- Eliminate energy demand associated with aerobic biological treatment
- Maximize energy production from organics in raw sewage
- Concentrate nutrients to facilitate subsequent recovery
- Reduce treatment plant footprint
- Simplify sewage treatment plant operation
- Facilitating mitigation of GHG emissions
- Produce high quality water suitable for many high value direct uses

Research Sponsors

(Utilities, Federal Government, Academia, Vendors, Professional Organizations)

P – Primary (one only),

S – Secondary (unlimited)

P – WERF (tailored collaboration with major motivated utility, etc.)

S – SBIR programs at Department of Energy, U.S. EPA, National Science Foundation, etc.

S – Electric Power Research Institute, WaterReuse Foundation,

Sequencing
Preceding Activities

Mitigating membrane fouling (new membranes or methods)

- Reduce membrane operating energy for sewage treatment
- Acceptance by regulators and utilities as alternative to aerobic treatment
- Pilot scale validation and membrane challenges
- LCA and LCC analyses
- Significant plant demonstration program

Subsequent Activities

- Full scale beta site plant operation
- Promote water industry acceptance (e.g., new technology risk, liability, etc.)
- Continued operational research for performance optimization

Duration __4___Years

Level of WERF Involvement ___TBD____%

Note that research may occur independently of WERF, requiring only a verification role for WERF.

A.11 Sustainability Metrics (vs. Wastewater Planning and Resource Recovery)

General Scope/Objectives

- Introduce the broad application best practices worldwide to provide objective metrics for Triple Bottom Line (TBL) analyses used in master planning of wastewater facilities. This will ensure that realistic environmental measures are included in planning of wastewater systems and provide objective input to TBL Analyses.
- LCA as tool for quantifying environmental impacts for TBL has remained in the university and has only impacted a single municipality's planning (Goteburg); it has high potential to support better environmental decision making and provide the metrics for the environmental leg of TBL.
- LCC analysis has had a lot of practical application but it can be more rigorously applied than in the past to capture more of the costs of a project and populate the cost leg of TBL analysis.
- The social leg of the TBL approach tends to under represented in terms of the full set of issues that must be considered in project implementation. Engineers tend not to address this leg very well. Tools under development represent current thinking –socio technological (technological includes environmental and economic) design and planning tools that integrate stakeholder involvement and interdisciplinary teams.

Benefits/Outcomes

- Use of common approaches for analyses that will be less subject to challenge and improved decision making in planning new infrastructure investments
- Benchmarking of internationally available tools and success or failure of outcomes in wastewater planning conducted with LCA, LCC, and social metrics

- Improved use of limited resources, lower costs, improve public acceptance, less environmental impacts

Research Sponsors

(Utilities, Federal Government, Academia, Vendors, Professional Organizations)

P – Primary (one only), S – Secondary (unlimited)

P – Utilities, National Governments, US State governments

S – Consultants

S – Academia

Sequencing
First Activities

- Bench mark best practices, develop case histories for major utilities employing LCA and LCC in master planning (e.g., Goteburg, Sydney)
- Review academic literature
- Review available LCA model platforms found useful in wastewater planning and resource recovery applications.
- Refine LCC experience including recent efforts to monetize environmental impacts as an alternative measure available with LCA approaches.
- Social metrics are difficult to realistically develop without broad stakeholder involvement; examine case histories where this has been successful and recommend improvements.
- Review the few applications of LCA and LCC that have entered into the master planning process, examine stakeholder involvement, assess causes of successes or failure in their application

Subsequent Activities

- Take initial efforts that have developed socio-technological design and planning tools that integrate all of the elements of sustainability, cost and social metrics and improve and apply them.
- "Test drive" the tools in new planning applications, learn and refine them

Duration 5 to 7 Years

Duration needed because of long lead times to determine success of planning and acceptance.

Level of WERF Involvement 25 %

This is likely an internal task leveraging the best of the tools available internationally to develop case histories, refine tools etc. An appropriate effort of the Global Water Research Coalition.

A.12 Enhanced Primary Treatment

General Scope/Objectives

- Improve primary settling for increased solids feed to digestion processes

- Better understanding of influent particle characteristics and flocculation in primary clarifiers
- Develop additives that are equally effective to current chemicals for coagulation and flocculation but add minimal inert material (reduce inert sludge quantity)
- Physical improvements for enhanced flocculation, coagulation;
- Use of modern tools such as cfd for evaluation/design of primary clarifiers
- As an alternate to primary clarification, examine greater application of anaerobic treatment of wastewater (e.g., application of UASB and other alternatives to more dilute and colder wastewater; prescreening/fixed film)
- Optimize methane collection from anaerobic pretreatment processes (aeration and gas capture, vacuum stripping)
- Recovery of dissolved methane formed in collection system or primary treatment before sending wastewater flow to secondary treatment
- Better understanding of collection system impacts on primary treatment (fermentation in sewers, etc.)
- Impact of collection system design on influent characteristics (aerated vs. fermented, what can be done in design to manage the outcome)
- Understand the tradeoffs between competing objectives (e.g., VFA production for bioP compared to maximizing energy recovery from biosolids)

Benefits/Outcomes

- Reduce organic loading to secondary treatment (improves energy balance)
- Reduced chemical consumption and insert solids production compared to conventional CEPT
- More volatile solids available for energy recovery
- Innovations to reduce potential energy loss within collection system

Research Sponsors

(Utilities, Federal Government, Academia, Vendors, Professional Organizations)

P – Primary (one only), S – Secondary (unlimited)

P – WERF

S – Utilities

S – Academia

S – Vendors (chemicals, equipment)

Sequencing

- Resurrect recommendations from WERF study on primary settling
- Frame the question – Identify and communicate limiting factors in primary settling and what could be improved – develop roadmap

Do the Below In Parallel

Examine/determine collection system impacts on primary settling and also related methane loss in sewers	Research physical improvements
Research/develop chemical additives	Determine downstream treatment following

	UASB
Determine alternatives to UASB systems for colder or more dilute wastewater (e.g., fixed film)	Develop white paper focused on identifying and quantifying the tradeoffs between competing objectives (e.g., VFA production for bioP compared to maximizing energy recovery from biosolids, others)

Subsequent Activities

Compile research results. Develop coordinated set of conclusions and guidance based on results of the individual research projects (i.e. clearly communicate the "so-what")

Duration __3 to 5__ years

Level of WERF Involvement __TBD__ %

APPENDIX B

"PLANT OF THE FUTURE" CONCEPT SUBMISSIONS

A variety of plant-wide concepts were submitted by some attendees following the workshop. The submissions as received are included below.

B.1 One Team's Vision of the Future for WWTPs

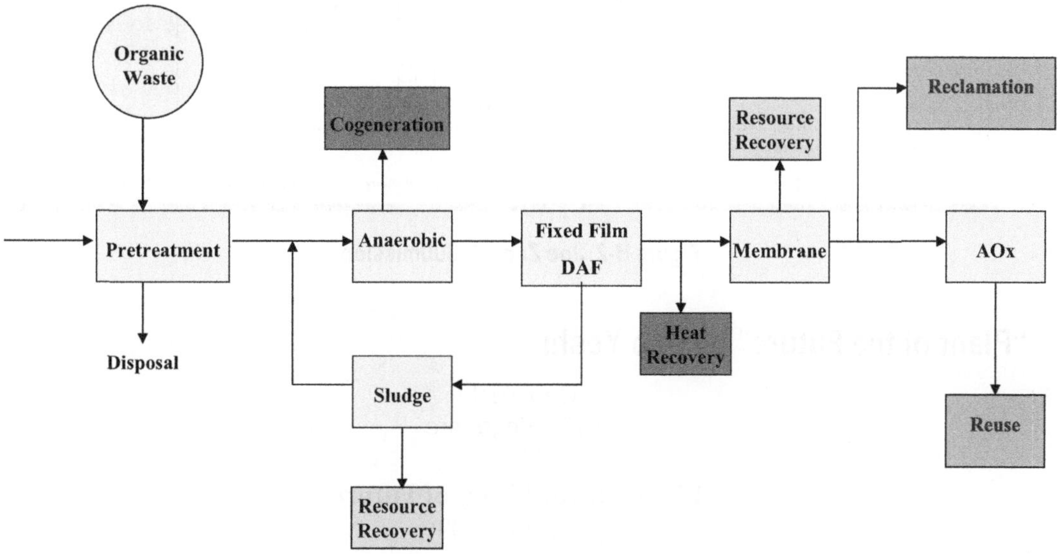

Figure B–1. One Team's Vision.

B.2 A Vision of the Future by Joe Zuback

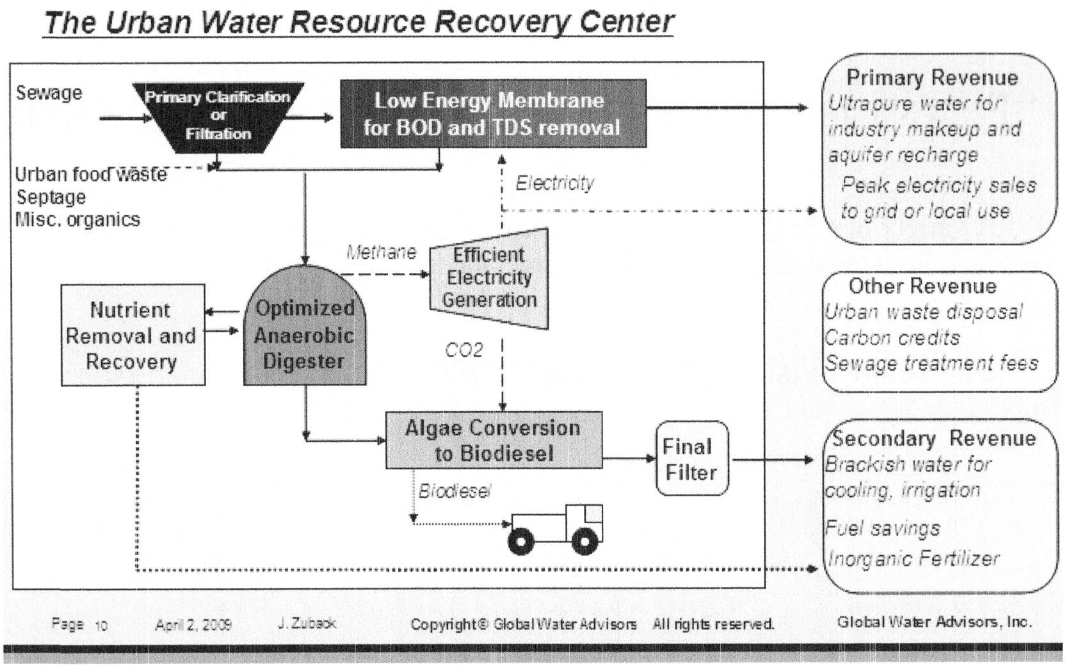

Figure B-2. Joe Zuback Submission.

B.3 "Plant of the Future" by Cao Yeshi

Cao Yeshi
PUB, Singapore

Note

Plants (liquid line) of Future

(5th June 2009)

B.3.1 Overall Concept

Aeration energy takes 40-60% of total energy use and it should be the limiting factor for energy saving and the focused point of the research in the time framework till 2030. Oxygen demand in the current liquid process derived from (1) COD biodegradation; (2) ammonia oxidation (nitrification); and (3) bio P-uptake. The aeration energy will be significantly reduced accordingly should the demands of three components could be reduced or eliminated.

Therefore, a new process, which adopts the anammox process into the mainstream municipal sewage treatment, is proposed as showed in Figure 12-3. The ideas behind the new process are:

- To maximize methane production in anaerobic digesters by retaining the soluble, colloid and particular COD prior them entering biological process to a maximum extent for maximizing energy generation in the solid line;
- To minimize aerobic heterotrophic COD conversion;
- To reduce/eliminate oxygen supply for nitrification in both main and side stream treatments and biological phosphorus removal; and

- To minimize excess sludge production through minimizing aerobic heterotrophic conversion and maximizing autotrophic anaerobic denitrification in the anammox process.

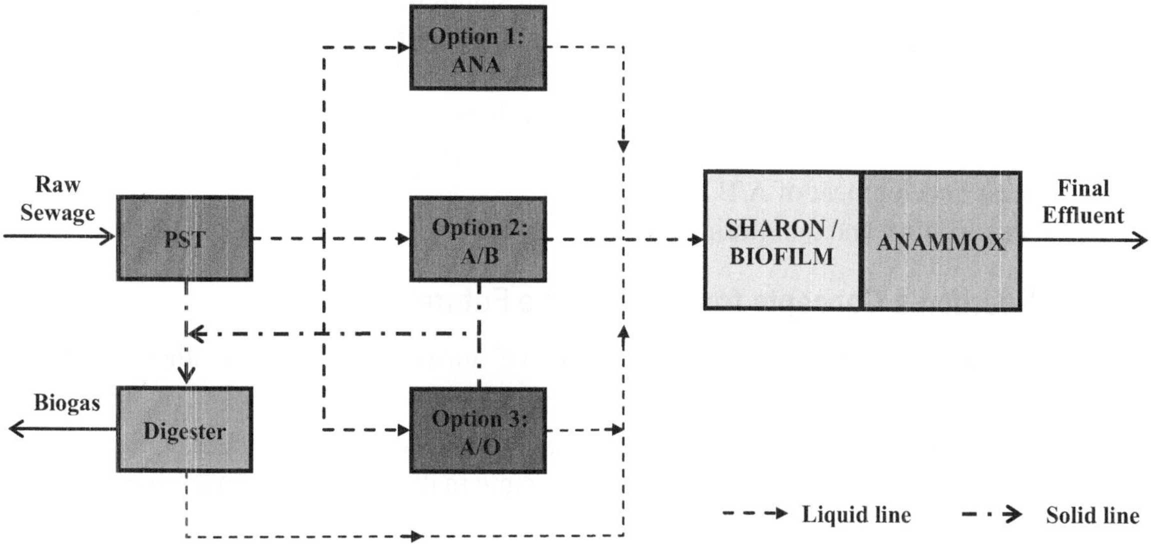

Figure B-3. Main Sub-processes in the New Process.

Three alterative sub-processes are considered to allocate following the primary setting tank and before the anammox process in the mainstream (Figure B-3).

- UASB reactor: a direct conversion of COD into methane, which makes omitting digester possible. However, methane collection in full-scale, which may be not so easy, methane dissolved into the liquid and high infrastructure cost are the major issues to be considered or solved.
- Both A/B and A/O are activated sludge processes but with very short HRT (2-4 h) and SRT (2-5 d) for maximum harvesting of COD (solids, colloid and soluble) and ammonia as much as possible with as little biological conversion (into CO_2) to have maximum COD mass to be sent to anaerobic digestion for methane production.
- The advantage of A/O process lies on possibly less aeration needed because of anaerobic stabilization. P could be recovered in the digesters. A quantitative comparison among the three alternatives and the interrelated relationship with the ANOMMOX and digestion is needed.

The effluent of the preceding biological process contains mainly ammonia and some soluble COD. Both will be treated in the major unit of anammox process in the mainstream. Ammonia could be partially oxidized to nitrite in a biofilm nitritation process with a low supply of aeration. Nitrite will then react with ammonia mainly through autotrophic denitrification under anaerobic condition in the anammox process.

The primary sludge from the primary settling tank and the secondary sludge from the biological process before the anammox process will be digested in the anaerobic digesters (or the UASB reactor) where methane is generated. The supernatant will then be treated together with the effluent from the biological process in a biofilm nitritation/anammox process.

B.3.2 Areas of Study

The major challenge is to develop anammox in the main stream rather than side stream i.e. under low ammonia mass loading and low temperature conditions.

- Enrichment of *Planctomycete*-like anaerobic ammonium-oxidizing bacteria for the ammonia concentration range between 30 and 50 mg N l^{-1} and tropical climate conditions (25–30°C) and temperate conditions (18–25°C)
- Immobilization of biomass: types of biomass immobilization such as using carrier materials, classic biofilm and granulation sludge etc
- Process development of A/B and A/O; and
- Process control and monitoring

B.4 Dave Reardon's Concepts for "Plant of the Future"

The following concepts apply to "inside the fence" ideas and not include others that could be done outside the fence like urine separation and in-pipe treatment.

1. Maximize performance of primary treatment to reduce loading on energy intensive secondary processes and divert as much organic matter directly to anaerobic digestion.
   - CFD modeling and baffling of conventional primary clarifiers to improve hydraulics, reduce short circuiting and create conditions for maximum suspended solids removal
   - Chemically enhanced primary treatment. Coagulation chemicals are an obvious drawback, but maybe we can do more with polymers and use less salts
   - Ballasted flocculation/sedimentation to replace primary clarifiers. Ballasted flocculation using sand or magnetic material can remove 95% of suspended solids vs. about 60% for conventional primary clarifiers
   - Screens/Filters such as the Salsnes technology
2. Reduce energy in secondary process.
   - Improve alpha. Alphas at the heavily loaded inlet end of a plug flow aeration tank are extremely low, perhaps as low as 0.2 for BOD/secondary plants. Plants are now designed to provide extra aeration at the precise location where oxygen transfer efficiencies are the lowest. We need to pay more attention to this. Several ideas could be considered. Mechanical aeration in the first cell (alphas for mechanical aerators are 0.7-0.9, even in heavily loaded cells). Coarse bubble diffusers have better alphas than fine bubble diffusers and may actually transfer oxygen more efficiently in heavily loaded aeration cells. Finally, step feed may have aeration advantages because there is no heavily loaded low alpha zone. The "plant of the future" will likely use nitrate from recycled mixed liquor to do the heavy lifting in the low alpha zones
   - Use more attached growth systems. Attached growth systems use a fraction of the energy of conventional suspended growth systems. The much maligned trickling filter/ biotower is a very low energy user but with obvious disadvantages related to nutrient removal and potential odors. Nevertheless, if a biotower is coupled with ballasted flocculation/sedimentation for primary and secondary clarifiers you would have a high performance secondary treatment system that produces a fraction of the secondary biosolids compared to conventional activated sludge

treatment and use substantially less energy. The attached growth concept could possibly reduce energy in activated sludge systems. The IFAS concept (either fixed or suspended growth media) may have energy advantage although I have not heard this claim from manufacturers nor seen data in the literature. Colloidal material from fixed growth systems could impact effluent water quality.

- Properly designed and more efficient blower systems. New high speed "turbo" type blowers are more efficient and are gaining popularity. The "plant of the future" will include them. More importantly, someday we may be able to train designers to "right size" blower equipment. Blower systems are normally grossly oversized causing severe inefficiencies when plants operate at average and low loading conditions. Providing flexibility in blower sizing is a low tech and highly effective method of reducing overall energy in activated sludge systems.
- Anaerobic secondary treatment. This could use a fraction of the energy (and produce a fraction of the biosolids) of aerobic systems but with its own set of issues including nutrient removal and methane removal.

3. **Preconditioning biosolids before anaerobic digestion and increasing biosolids feed concentrations to digesters.** Acid phase digestion technology is applied extensively now. Thermal hydrolysis (such as CAMBI) is gaining more attention but energy balances must be calculated to see if overall plant energy is improved using this type of process. "Crapper zappers" and mechanical shearing have been tried with marginal success. The "plant of the future" will employ improved, energy efficient methods to hydrolyze biosolids and destroy cell walls of activated sludge, thus improving aerobic digester vs. destruction, increasing methane production, and improving dewaterability.
4. **Improved cogeneration performance.** We will see improvement in efficiency of cogeneration equipment. Present cogeneration equipment has a mechanical efficiency of no more than 32–34%. New engines are now available that have mechanical efficiencies of nearly 40%. Fuel cells are better yet, with claimed efficiencies approaching 50%. If the cost of fuel cells can be reduced, they will be a contender for cogeneration in the future(one of their biggest advantages is low emissions)
5. **More energy efficient disinfection.** UV disinfection is all the rage. Unfortunately the energy penalty is high, with UV disinfection using as much as 10–20% of plant power. Future plants that use UV disinfection will have a more energy efficient UV light source and be able to "turn down" effectively so that UV power matches flow. Alternatively, improved suspended solids removals with filtration or membranes may allow alternative chemicals to chlorine for disinfection. There are many possibilities including peracetic acid.
6. **Energy management systems.** The "plant of the future" will use a SCADA based energy management system to monitor and control all energy in the plant. Energy data will be available real time especially demand. Energy consumption metrics will be used as a diagnostic tool.
7. **Attention to sub-process energy consumption.** Systems like channel aeration and plant recycled water are always overdesigned with little attention paid to turn down or operation at non peak conditions. The aggregate energy consumption of miscellaneous systems can be as high as 10–20% of total plant power. The "plant of the future" will have capability for turndown and optimization of these systems to reduce energy consumption.

8. **Import of FOG, food waste and septage to augment anaerobic digester gas production.** These techniques, used sporadically now, will be commonplace in the future at plants with anaerobic digesters
9. **Increased use of biosolids incineration.** This eliminates anaerobic digesters and related energy consumption. Fluidized bed incinerators will be easier to permit. Ash can be used as a concrete admixture. Energy and cost related to biosolids disposal are greatly reduced. Waste heat can generate steam for steam turbine electricity generation.
10. **Increased use of green energy.** We will see increased use of photovoltaic and wind energy production, either on or off site. POTWs will be able to claim energy neutral (sustainable) plants by optimizing energy inside the fence and claiming credit for green energy.
11. **Improved involvement of management, engineering and O&M staff.** Energy optimization is not just about the latest technology; it's really about the attitude of the staff. Staff for the "plant of the future" will have energy training, energy consumption data, and an energy management system to monitor and control processes. In addition, the plant will have and energy team and energy champion that meet regularly to discuss optimization ideas and implementation. These plants will also have an incentive program (money is not the only available incentive) to enhance the profile of energy conservation.

B.5 Contributions from Mark van Loosdrecht from Published Literature

Pergamon

Wat. Sci. Tech. Vol. 35, No. 9, pp. 171–180, 1997.

Printed in Great Britain
0273–1223/97 $17·00 + 0·00

PII: S0273-1223(97)00195-9

TOWARDS A MORE SUSTAINABLE MUNICIPAL WASTEWATER TREATMENT SYSTEM

Mike S. M. Jetten*, Svein J. Horn** and
Mark C. M. van Loosdrecht*

* *Kluyverlaboratorium for Biotechnology, Faculty of Chemical Technology and Material Sciences, Delft University of Technology, Julianalaan 67, NL 2828 BC Delft, The Netherlands*
** *Department of Biotechnology, Faculty of Chemistry and Chemical Technology, Norwegian University of Science and Technology, N-7034 Trondheim, Norway*

ABSTRACT

The present day wastewater treatment practise can be significantly improved when new microbial processes are introduced. In this paper a new concept is discussed along general lines. The concept consists of a first stage in which a maximal sludge production is achieved. In this way COD is removed as sludge and aeration requirements minimized. This sludge can be digested to yield methane for energy generation. The nitrogen removal is based on a partial oxidation to nitrite and subsequent denitrification with ammonium as electron donor. © 1997 IAWQ. Published by Elsevier Science Ltd

Figure B-4. Paper #1 Abstract.

Integration of Processes to Treat Wastewater and Source-Separated Urine

Jac A. Wilsenach[1] and Mark C. M. van Loosdrecht[2]

Abstract: Human urine contributes 80% of the total nitrogen and 40–50% of the total phosphate load to municipal wastewater. This study examines the impact of separate urine collection and treatment on wastewater treatment. An integrated wastewater and urine treatment process was defined, in which single high-rate ammonium removal over nitrite and anaerobic ammonium oxidation processes and struvite recovery are at the heart of the nutrient management. The model study demonstrated that if 50% or more of urine were collected and treated separately, integrated wastewater treatment with more compact and energy-efficient processes would be possible. The integrated wastewater and urine treatment is compared to an existing state-of-the-art treatment process. The main advantage of urine separation is not only a better effluent quality. Existing processes including tertiary treatment can already produce very good effluent quality with total effluent nitrogen and phosphate concentrations of 2.5 and 0.5 g/m^3, respectively. The main advantage of urine separation is the production of this same good effluent quality with a remarkable saving in resources. With sufficient urine separation, generation of net primary energy is possible.

Journal of Environmental Engineering ASCE March 2006 page 331

Figure B-5. Paper #2 Abstract.

APPENDIX C

RAW NOTES FROM BREAKOUT SESSIONS

For the sake of completeness, the raw notes created as deliverables by the various breakouts are recorded below. These notes were presented to the workshop as a whole by each breakout group, and from these presentations the above prioritized lists were created.

C.1 Group A

What effluent/reuse quality will be needed in 2040?

C.1.1 Designated use based goals/limits

Quality types

Water

- Chemical Standards
  - Microconstituents
- Biological Standards
  - Bacteria/virus resistance
- Physical Standards
  - Nanoparticles
  - TDS
  - Temperature

Solids

- Beneficial use/zero disposal
  - Sequestration
  - Complete reuse
- Minimization
  - Energy recovery

Medium and Product Management

Water

- Algae
- Thermal energy recovery
- Aquaculture

Air

- GHG control
- Odor
- Only biogenic carbon emission
- Energy output for C use

What will be the Common Treatment Technologies in 2040?
Better separation and reuse processes for:
Water

- Nutrients
- Thermal/Heat
- Organic Carbon
- Inert solids (grit)
- FOGs
  - Reduce or eliminate forced aeration (blowers and diffusers)
  - Membrane technology
  - Heat pumps
  - Anaerobic Treatment
  - Localized liquid treatment, centralized solids management
  - Rheology modification
  - More efficient thermal oxidation
  - Better biogas treatment technology
  - Need technologies to make products of grit and FOGs

How can we Achieve Water/Carbon/Energy Neutral Treatment?

- More efficient heat and energy reuse
- Hydraulic energy recovery
- Biomass production (targeted)
- Emerging technologies
  - Nanomaterials
  - Biotechnology
- Geological sequestration of sludge
- Water conservation practice
- Collection sys mgmt
  - Reduce I/I
- Complete residuals reuse
- Integrated regional management of resources
  - Water, C, N, P
- Trading systems?
- Source separation/reduction of pollutants
- Decentralized treatment
- Integration with natural treatment systems

C.2 Group B

What effluent/reuse quality will be needed in 2040?
Water

- Pathogen virulence/resistance
- Individual pathogens/regulations, move away from indicators
- Metals at very low levels (e.g., Hg)
- Nanoparticle standards

Air

- Local air pollution regulations
- Bio-aerosols
- Stricter odor control (decentralized)

Residuals

- Transportation emissions

What will be the common treatment technologies in 2040?

- Energy recovery from collections systems
- Greater use of anaerobic tech
  - Including sulfur management
- Anaerobic technologies in the collection system
  - Sulfur/corrosion
- Collection systems that minimize I/I
- Collection and treatment of off-gases
- Materials that reduce pumping (head) losses
- Food-grinders revisited
  - Increase use to enhance biogas
- Maximize carbon removal in primary
- Microbial-fuel cells commercialized for decentralized systems
- Better technologies for selection and removal of inorganics from mixed liquid
- Thermal and hydrothermal WAO supercritical water oxidation
- N/P recovery technologies
  - Source separation of waste
- Systemized instrumentation and control
- Autotrophic and phototrophic nutrient removal tech
- Multiple barrier disinfection to reduce
  - Pathogeneity
  - Resistance
  - Virulence

How can we Achieve Water/Carbon/Energy Neutral Treatment?

- Balancing water/energy/carbon
- Developing system-wide decision support tools for accounting for water/energy/C
- Regional/holistic boundaries
- Systems approach to Water/Eng/C/Nutrients
- Unified tools for calculating water footprint, C footprint, energy footprint, human footprint
- Decision support tool, TBL: economic, environmental, social

C.3 Group C

What effluent/reuse quality will be needed in 2040?
Effluent Quality

- Discharge
- Impact based TMDL and public health (pathogens and contaminants)

Reuse

- Use application based

Biosolids

- Zero waste discharge
- Biosolids to energy
- Biosolids to marketable products

Air

- Zero odor facility
- Aerosols
- Particulates
- NO_X, SO_X, Hg
- GHG per person, CO_2, N_2O, CH_4

What will be the common treatment technologies in 2040?

- Enhanced concentration of raw WW organic matter
- Anaerobic liquid treatment/methane recovery
- Inorganic mining – chemical and biological
  - N&P, metals, and other materials
- Immobilized enzyme treatment
- Improved gas treatment (H_2S, siloxanes)
- Carbon storage (organic and inorganic)
- Capture and reuse of off-gases within treatment (improved efficiency)
- Fixed films
- Media filtration

How can we Achieve Water/Carbon/Energy Neutral Treatment?

- Smart grid technology
- Satellite/size energy vs. water
- Remove water from biosolids
- Improved solid-liquid separation (primary)
- Hydrolysis of solids – enhanced methanogenesis
- Energy recovery – heat hydraulic, carbon, solar, wind
- Additional carbon sources from energy recovery
- Real-time energy management system at plant
- Optimization of operations (incentive-based)
- Flexible design for operation capacity
- Improved electricity production from biogas/biosolids

- Implementation of existing advanced technologies for energy production/use and proper operations

C.4 Group D

What effluent/reuse quality needed for 2040?

Water and biosolids will be a valuable resource – the scenarios:

- Very high quality water for industrial use and potable use
- Surface water discharge – ecologically neutral or enhancing
- Marketable products within area of production (biosolids)
- Appropriate quality for agriculture other nonpotable reuse

What will be the common treatment technologies in 2040?

Very high Quality Water

Cheap efficient membrane technology with pretreatment/primary and purely physical/chemical separation. Concentrates to anearobic digester. Methane produced provides for process, CO_2 cycled to algae contactor. Centrate goes to P and N recovery then its effluent goes to photosynthetic treatment (N and P removed) and effluent goes to a user such to evaporative cooling with concentration of brine. Algae goes to biofuels, residuals to digester. Net energy producer claimed.

Surface water discharge for ecological enhancement

Diversion out of river: Polishing Wetlands for conventional plants and other upstream NPS.

Agriculture or other nonpotable uses

Primary treatment/pretreatment to increase primary sludges for digestion, Followed by oxidation ponds, possibly algae harvesting just for methane recovery

Biosolids

Low energy precondition of solids to maximize release of biodegradable materials and maximize biogas production

How can we achieve carbon neutral and energy-neutral treatment?

- Energy management systems – everything in the plant and outside the plant that we can effect
- Develop microbial fuel cells and MEC technology to field scale reality
- Maximize input of degradable organics to anaerobic processes
- Maximize energy value (outside the fence value)
- Use biosolids to maximum carbon sequestration in the environment (externality)
- Public education or we won't get there – recycle locally – e.g., the City of Altamont Springs in Florida. Deal with the "yuck factor" directly.
- Heat recovery from effluent, outfall energy recovery, water conservation

General comments in plenary discussion

- Just seeing the beginning bio technologies, treatment oriented super-bugs
- Using expertise outside WWT arena, "crossover" technologies

- Improve metrics, TBL is qualitative, “sustainable” should be viewed as a tool not an end
- Decision support systems (methodologies)
- Public education
- Rebranding
- Risk management, complicated, important, risk has many meanings

C.5 Thursday Breakouts

In the following breakout listings, two additional pieces of information are provided with several items. These relate to the Roadmap Concept Rating scale prioritization category (Roman Numeral, in red) and the consensus (in blue, refers to other groups that identified the same item0. The prioritization category is based on the following table:

Roadmap Concept Rating Scale (I – IV)

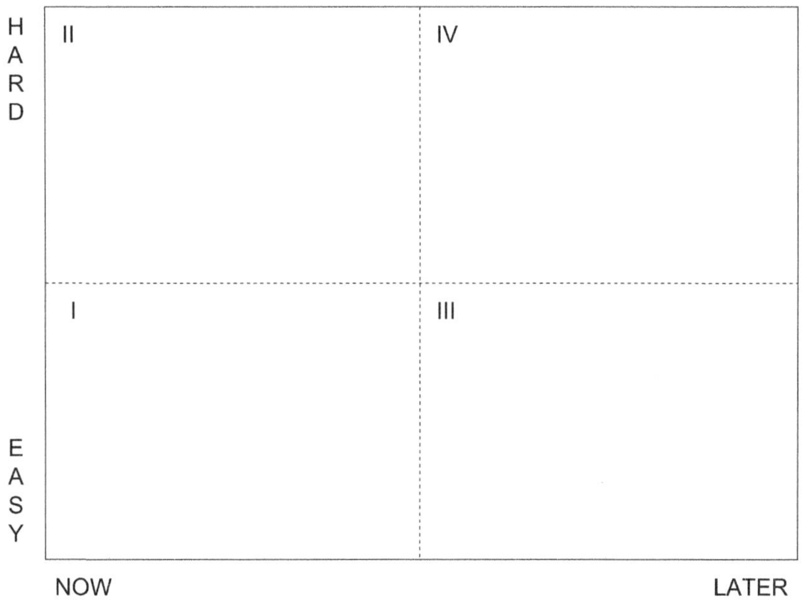

9.1.2 Group A

Equipment

- Biogas cleaning I (B)
- Improved aeration equipment II (B, C, D)
- Selective sorption media IV
- Microbial fuel cells IV (D)
- Heat pumps I
- Non-fowling membrane (all around better) II (B, D)

Treatment Processes

- Anaerobic treatment I (B, C, D)
- Thermal recovery III (B, C)
- Solids rheology/hydrolysis II
- Fixed film processes I (C, D)

- N&P recovery I (B, C)
- Photosynthesis/algae IV (B)
- Enzyme treatment IV

Management

- Comprehensive decision support tools to balance C/water/energy I (B, D)
- Regional solids co-treatment II
- Centralized solids management w/scalping III (C)
- Energy management systems I
- Carbon/nutrient trading/ regional management IV
- Minimize I/I in collection system I (B)

9.1.3 Group B

Equipment

- Piece of equipment to improve primary treatment (e.g., coarse membrane) II
- Energy – efficient equipment I (A, C, D)
- Real – time sensors/automation that allow us to employ more complex technologies and processes (system level – plant, collection, stormwater, etc) I, III
- Biogas cleaning/processing equipment I (A)
- Membrane technologies IV (A, D)
  - Low energy
  - Anti-fouling/low cost
- Heat Recovery equipment I (A, C)
  - Sewage, hot gases, effluent, etc.
- I/I prevention equipment II (A)

Treatment Processes

- Maximize conversion of WW carbon to methane II, IV (A)
  - Low temp anaerobic treatment
  - High temp " "
  - Digestion route IV
  - Enhance primary treatment to remove C early in the process
- Auto/phototrophic nutrient removal of treated waters ("reject water") II, IV (A, C)
- Thermal & Hydrothermal processing of sludge that reduces C footprint and approaches energy neutral II (A)
- Low energy micro pollutant treatment processes (AOP, biofilms, sorption, RO) IV (A)
- Membrane application in combination with other processes above (A, D)
- Nutrient recovery technologies I (A, C)
- Advanced disinfection II (D)
- GHG minimization technologies II
- Biofilters (GHG & odors) I

Management

- Smart control III (D)
  - Better use of operator knowledge
  - I/C output

- Decision support systems
- Socio-technological/design & planning methodology II (A, D)
  - Consider all (energy, water, C) users
  - Stakeholder inclusion and engagement
  - Iterative process
  - Get it working in the field (highly interdisciplinary)
- Education II (D)
  - Public
  - Future water professionals
  - Regulators, policy makers
  - Other stakeholders
- Sustainability coordinator (czarina) I

9.1.4 Group C

Equipment

- Better aeration equipment I (A, B, D)
- Collection system heat recovery IV
- Thermal biosolids processing III (B)
- Media filtration I

Treatment Processes

- Anaerobic treatment (liquid) IV (A, B, D)
- Heat recovery IV (A, B)
- Better disinfection (non Cl_2) II (B, D)
- Fixed films I (A, D)
- Co-generation I
- Biogas treatment II (A)
- Inorganic mining IV (D)

Management

- Co-proc of food wastes I
- Subsurface residuals injection IV
- N & P recovery I (A, B)
- Decentralized liquid and centralized solids III (A)
- Public communication II (B, D)
- Water conservation II
- Green energy I

9.1.5 Group D

Equipment

- Optimize existing
- Optimize all things membranes II (A, B)
- Aeration equip optimize, near term, can all be furthered I (A, B, C)
- Better sensors I
- *New*
- Fixed film processes optimized IV (A)
- microbial fuel cells IV (A)

Treatment Processes

- Optimize existing
- Enhanced primary (e.g., flocculation, coagulation, anaerobic integration) II (A, B)
- Fixed film, better understand operation I (A)
- Optimization anaerobic systems (solids handling, maximize recovery, volatile) II (A, B, C)

New

- MEC and microbial fuel cell development IV (A)
- Pretreatment requirements for advanced oxidation practices to meet low EPOC from process perspective (e.g., w/o membranes or filter) IV (B)
- Processes for selective recovery of products, long term (metals, plastics) IV (A)
- Maximize energy yield of carbon treatment IV (B)
- "Friendly" disinfectant (e.g., risk, energy, no disinfection byproducts) IV (B)
  - Low C footprint advanced disinfection

Management

- Optimize existing
- Identify best management practices (design and O&M) for optimum operating point I (A, B)
- Public education (pop culture, wiki, etc) II (B)
- Marketing products II
- Means for trusted information sharing and communication and technology evaluation (Global Water Research Coalition) II
- Rewriting and de-bottlenecking I

New

- LCA, LCC best practice tools IV (B)
- DSS with controls, support for real-time operational support III (A, B)
- Benchmark of practices (all industries) staff and management education III

References

Crawford, G.V. (2010) *Best Practices for Sustainable Wastewater Treatment,* Report no. OWSO4R07a; Alexandria, VA: Water Environment Research Foundation.

Jetten, M.S.M., S.J. Horn, and M.C.M. van Loosdrecht (1997) "Towards a More Sustainable Municipal Wastewater Treatment System," *Water Sci. Technol.* Vol. 35. No. 9. pp.171–180.

Kalogo, Y. and H. Monteith (2008) *State of Science Report: Energy and Resource Recovery from Sludge,* Report no. OWSO3R07; Alexandria, VA: Water Environment Research Foundation.

Shizas, I. and D.M. Bagley (2004) "Experimental Determination of Energy Content of Unknown Organics in Municipal Wastewater Systems," *J. Energy Engin.* 130, 2, pp. 45-53.

Tchobanoglous, G., H. Leverenz, and P. Gikas (2009) "Impacts of New Concepts and Technology on the Energy Sustainability of Wastewater Treatment" Presented at *Climate Change, Sustainable Development and Renewable Energy Sources* conference at Aristotle University of Thessaloniki, Greece, October 15–17, 2009.

U.S. Environmental Protection Agency (2008) *Clean Watershed Needs Survey 2004 Report to Congress.* Washington, DC, U.S. EPA.

WASTEWATER UTILITY

Alabama
Montgomery Water Works & Sanitary Sewer Board

Alaska
Anchorage Water & Wastewater Utility

Arizona
Avondale, City of
Glendale, City of, Utilities Department
Mesa, City of
Peoria, City of
Phoenix Water Services Dept.
Pima County Wastewater Management
Safford, City of
Tempe, City of

Arkansas
Little Rock Wastewater Utility

California
Central Contra Costa Sanitary District
Corona, City of
Crestline Sanitation District
Delta Diablo Sanitation District
Dublin San Ramon Services District
East Bay Dischargers Authority
East Bay Municipal Utility District
El Dorado Irrigation District
Fairfield-Suisun Sewer District
Fresno Department of Public Utilities
Inland Empire Utilities Agency
Irvine Ranch Water District
Las Gallinas Valley Sanitary District
Las Virgenes Municipal Water District
Livermore, City of
Los Angeles, City of
Los Angeles County, Sanitation Districts of
Napa Sanitation District
Novato Sanitary District
Orange County Sanitation District
Palo Alto, City of
Riverside, City of
Sacramento Regional County Sanitation District
San Diego Metropolitan Wastewater Department, City of
San Francisco, City & County of
San Jose, City of
Santa Barbara, City of
Santa Cruz, City of
Santa Rosa, City of
South Bayside System Authority
South Coast Water District
South Orange County Wastewater Authority
South Tahoe Public Utility District
Stege Sanitary District
Sunnyvale, City of
Union Sanitary District
West Valley Sanitation District

Colorado
Aurora, City of
Boulder, City of
Greeley, City of
Littleton/Englewood Water Pollution Control Plant
Metro Wastewater Reclamation District, Denver

Connecticut
Greater New Haven WPCA
Stamford, City of

District of Columbia
District of Columbia Water & Sewer Authority

Florida
Broward, County of
Fort Lauderdale, City of
Jacksonville Electric Authority (JEA)
Miami-Dade Water & Sewer Authority
Orange County Utilities Department
Pinellas, County of
Reedy Creek Improvement District
Seminole County Environmental Services
St. Petersburg, City of
Tallahassee, City of
Toho Water Authority
West Palm Beach, City of

Georgia
Atlanta Department of Watershed Management
Augusta, City of
Clayton County Water Authority
Cobb County Water System
Columbus Water Works
Fulton County
Gwinnett County Department of Public Utilities
Savannah, City of

Hawaii
Honolulu, City & County of

Idaho
Boise, City of

Illinois
Decatur, Sanitary District of
Greater Peoria Sanitary District
Kankakee River Metropolitan Agency
Metropolitan Water Reclamation District of Greater Chicago
Wheaton Sanitary District

Indiana
Jeffersonville, City of

Iowa
Ames, City of
Cedar Rapids Wastewater Facility
Des Moines, City of
Iowa City

Kansas
Johnson County Wastewater
Unified Government of Wyandotte County/ Kansas City, City of

Kentucky
Louisville & Jefferson County Metropolitan Sewer District
Sanitation District No. 1

Louisiana
Sewerage & Water Board of New Orleans

Maine
Bangor, City of
Portland Water District

Maryland
Anne Arundel County Bureau of Utility Operations
Howard County Bureau of Utilities
Washington Suburban Sanitary Commission

Massachusetts
Boston Water & Sewer Commission
Massachusetts Water Resources Authority (MWRA)
Upper Blackstone Water Pollution Abatement District

Michigan
Ann Arbor, City of
Detroit, City of
Holland Board of Public Works
Saginaw, City of
Wayne County Department of Environment
Wyoming, City of

Minnesota
Rochester, City of
Western Lake Superior Sanitary District

Missouri
Independence, City of
Kansas City Missouri Water Services Department
Little Blue Valley Sewer District
Metropolitan St. Louis Sewer District

Nebraska
Lincoln Wastewater & Solid Waste System

Nevada
Henderson, City of
Las Vegas, City of
Reno, City of

New Jersey
Bergen County Utilities Authority
Ocean County Utilities Authority

New York
New York City Department of Environmental Protection

North Carolina
Charlotte/Mecklenburg Utilities
Durham, City of
Metropolitan Sewerage District of Buncombe County
Orange Water & Sewer Authority
University of North Carolina, Chapel Hill

Ohio
Akron, City of
Butler County Department of Environmental Services
Columbus, City of
Metropolitan Sewer District of Greater Cincinnati
Montgomery, County of
Northeast Ohio Regional Sewer District
Summit, County of

Oklahoma
Oklahoma City Water & Wastewater Utility Department
Tulsa, City of

Oregon
Albany, City of
Clean Water Services
Eugene, City of
Gresham, City of
Portland, City of Bureau of Environmental Services
Lake Oswego, City of
Oak Lodge Sanitary District
Water Environment Services

Pennsylvania
Hemlock Municipal Sewer Cooperative (HMSC)
Philadelphia, City of
University Area Joint Authority

South Carolina
Charleston Water System
Mount Pleasant Waterworks & Sewer Commission
Spartanburg Water

Tennessee
Cleveland Utilities
Murfreesboro Water & Sewer Department
Nashville Metro Water Services

Texas
Austin, City of
Dallas Water Utilities
Denton, City of
El Paso Water Utilities

WERF SUBSCRIBERS

Fort Worth, City of
Houston, City of
San Antonio Water System
Trinity River Authority

Utah

Salt Lake City Corporation

Virginia

Alexandria Sanitation Authority
Arlington, County of
Fairfax, County of
Hampton Roads Sanitation District
Hanover, County of
Henrico, County of
Hopewell Regional Wastewater Treatment Facility
Loudoun Water
Lynchburg Regional Wastewater Treatment Plant
Prince William County Service Authority
Richmond, City of
Rivanna Water & Sewer Authority

Washington

Everett, City of
King County Department of Natural Resources
Seattle Public Utilities
Sunnyside, Port of
Yakima, City of

Wisconsin

Green Bay Metro Sewerage District
Kenosha Water Utility
Madison Metropolitan Sewerage District
Milwaukee Metropolitan Sewerage District
Racine, City of
Sheboygan Regional Wastewater Treatment
Wausau Water Works

Water Services Association of Australia

ACTEW Corporation
Barwon Water
Central Highlands Water
City West Water
Coliban Water Corporation
Cradle Mountain Water
Gippsland Water
Gladstone Area Water Board
Gold Coast Water
Gosford City Council
Hunter Water Corporation
Logan Water
Melbourne Water
Moreton Bay Water
Onstream
Power & Water Corporation
Queensland Urban Utilities
SEQ Water
South Australia Water Corporation
Sunshine Coast Water
Sydney Catchment Authority
Sydney Water
Unity Water
Wannon Regional Water Corporation
Watercare Services Limited (NZ)
Water Corporation
Western Water
Yarra Valley Water

Canada

Edmonton, City of/Edmonton Waste Management Centre of Excellence
Lethbridge, City of
Regina, City of, Saskatchewan
Toronto, City of, Ontario
Winnipeg, City of, Manitoba

STORMWATER UTILITY

California

Fresno Metropolitan Flood Control District
Los Angeles, City of, Department of Public Works
Monterey, City of
San Francisco, City & County of
Santa Rosa, City of
Sunnyvale, City of

Colorado

Aurora, City of
Boulder, City of

Florida

Orlando, City of

Iowa

Cedar Rapids Wastewater Facility
Des Moines, City of

Kansas

Lenexa, City of
Overland Park, City of

Kentucky

Louisville & Jefferson County Metropolitan Sewer District

Maine

Portland Water District

North Carolina

Charlotte, City of, Stormwater Services

Pennsylvania

Philadelphia, City of

Tennessee

Chattanooga Stormwater Management

Texas

Harris County Flood Control District, Texas

Washington

Bellevue Utilities Department
Seattle Public Utilities

STATE

Connecticut Department of Environmental Protection
Kansas Department of Health & Environment
New England Interstate Water Pollution Control Commission (NEIWPCC)
Ohio Environmental Protection Agency
Ohio River Valley Sanitation Commission
Urban Drainage & Flood Control District, CO

CORPORATE

ADS LLC
Advanced Data Mining International
AECOM
Alan Plummer & Associates
Alpine Technology Inc.
Aqua-Aerobic Systems Inc.
Aquateam–Norwegian Water Technology Centre A/S
ARCADIS
Associated Engineering
Bernardin Lochmueller & Associates
Black & Veatch
Blue Water Technologies, Inc.
Brown & Caldwell
Burgess & Niple, Ltd.
Burns & McDonnell
CABE Associates Inc.
The Cadmus Group
Camp Dresser & McKee Inc.
Carollo Engineers Inc.
Carpenter Environmental Associates Inc.
CET Engineering Services
CH2M HILL
CRA Infrastructure & Engineering
CONTECH Stormwater Solutions
D&B/Guarino Engineers, LLC
Damon S. Williams Associates, LLC
Ecovation
EMA Inc.
Environmental Operating Solutions, Inc.
Environ International Corporation
Fay, Spofford, & Thorndike Inc.
Freese & Nichols, Inc.
ftn Associates Inc.
Gannett Fleming Inc.
Garden & Associates, Ltd.
Geosyntec Consultants
GHD Inc.
Global Water Associates
Greeley and Hansen LLC
Hazen & Sawyer, P.C.
HDR Engineering Inc.
HNTB Corporation
Hydromantis Inc.
HydroQual Inc.
Infilco Degremont Inc.
Jason Consultants LLC Inc.
Jordan, Jones, & Goulding Inc.
KCI Technologies Inc.
Kelly & Weaver, P.C.
Kennedy/Jenks Consultants
Larry Walker Associates
LimnoTech Inc.
Lombardo Associates, Inc.
The Low Impact Development Center Inc.
Malcolm Pirnie Inc.
Material Matters, Inc.
McKim & Creed
MWH
NTL Alaska, Inc.
O'Brien & Gere Engineers Inc.
Odor & Corrosion Technology Consultants Inc.
Parametrix Inc.
Parsons
Post, Buckley, Schuh & Jernigan
Praxair, Inc.
RMC Water & Environment
Ross & Associates Ltd.
SAIC
Siemens Water Technologies
The Soap & Detergent Association
Smith & Loveless, Inc.
Southeast Environmental Engineering, LLC
Stone Environmental Inc.
Stratus Consulting Inc.
Synagro Technologies Inc.
Tetra Tech Inc.
Trojan Technologies Inc.
Trussell Technologies, Inc.
URS Corporation
Wallingford Software
Westin Engineering Inc.
Wright Water Engineers
Zoeller Pump Company

INDUSTRY

American Electric Power
American Water
Anglian Water Services, Ltd.
Chevron Energy Technology
The Coca-Cola Company
Dow Chemical Company
DuPont Company
Eastman Chemical Company
Eli Lilly & Company
InsinkErator
Johnson & Johnson
Merck & Company Inc.
Procter & Gamble Company
Suez Environnment
United Utilities North West (UUNW)
United Water Services LLC
Veolia Water North America

Note: List as of 4/20/10

WERF Board of Directors

WERF Research Council

IWA Publishing's authorised EU representative for General Product Safety Regulations is Diane D'Arras, 15 rue Duret, 75116 Paris, France, e-mail: safety@iwap.co.uk.

Printed and bound by CPI Group (UK) Ltd, Croydon, CR0 4YY
13/05/2026
02109828-0007